The Ignite Project

Niyousha Raeesinejad •
Yousef Mehrdad Bibalan •
Mohammad Moshirpour

The Ignite Project

A Journey in Scrum

 Springer

Niyousha Raeesinejad
Department of Electrical
and Software Engineering
Schulich School of Engineering
University of Calgary
Calgary, AB, Canada

Yousef Mehrdad Bibalan
Department of Electrical
and Software Engineering
Schulich School of Engineering
University of Calgary
Calgary, AB, Canada

Mohammad Moshirpour
Department of Informatics
Donald Bren School of Information
and Computer Sciences
University of California, Irvine
Irvine, CA, USA

ISBN 978-981-19-4803-9   ISBN 978-981-19-4804-6   (eBook)
https://doi.org/10.1007/978-981-19-4804-6

This Springer imprint is published by the registered company Springer Nature Singapore Pte Ltd.
The registered company address is: 152 Beach Road, #21-01/04 Gateway East, Singapore 189721, Singapore

Preface

This guidebook tells the story of a software engineering internship project facilitated by the Schulich School of Engineering at the University of Calgary (U of C). The Agile team, composed of student developers pursuing either a master's or bachelor's degree in software engineering (SE), is led by a consulting tech company serving as the product owner, a graduate student, and a seasoned industry expert as project leads. The story is told from the perspective of an undergraduate SE student observing from the outside, as well as the perspectives of the Agile team members through meetings and retrospections. The book is intended for readers involved in research, education, and industry who wish to experience the same process from different angles and gain insight into tailoring different frameworks and training approaches to software projects.

In addition to telling the story of the software development process, this book also highlights the lessons learned by the interns and their progress in the industry-academia cross project as part of a master's software project course. The authors aim to show the preparation process for a SE internship, enhance students' knowledge of industry-relevant practices, and spread awareness to SE educators regarding the significance of adequately preparing students for the workforce by simulating more projects in classrooms. The book includes insights from a professor and two teaching assistants who served as the product owner and project leads, respectively.

Furthermore, this book serves as a testament to the economic impact of the Master of Engineering (MEng) Program in Software Engineering at Schulich, which has helped many interns secure software-related jobs following graduation. The authors wish to provide valuable advice to industry professionals in terms of understanding how to adequately support interns under constraints like time, cost, and effort. Ultimately, their long-term goal is to contribute towards further collaboration and foster a deeper relationship between academia and industry.

Biography

Author 1: Niyousha Raeesinejad

Niyousha is a software engineer with a strong and diverse academic background that includes a bachelor's degree in software engineering, a minor in Management and Society, and a certificate in Entrepreneurial Thinking, all earned from the University of Calgary (U of C). Niyousha is a research associate at the Software Engineering Practice and Education (SEPE) Research Group and has published multiple papers on software engineering education. In addition to her academic and research accomplishments, Niyousha has contributed several years towards fostering the tech innovation ecosystem for students as one of the past Presidents of Tech Start UCalgary, a student-driven club focused on software development and entrepreneurship at the UofC. She is a proud recipient of the 2022 Women in Stem Scholarship and actively advocates for her fellow peers pursuing careers in STEM. Niyousha's research interests include software engineering education, software innovation, hackathons, and open-source software.

Author 2: Yousef Mehrdad Bibalan

Yousef is a Ph.D. candidate at the Schulich School of Engineering at the University of Calgary. His career of over 20 years has involved combining the technical aspects of software development with a focus on human interaction, innovation, and continuous learning. Yousef has worked as a software engineer, instructor, trainer, coach, and author, honing his skills and expertise. His research is focused on using process mining to enhance software development processes by combining his expertise in software engineering practices, statistical methods, and machine learning. He aims to explore new insights and perspectives for process improvement by integrating traditional knowledge with cutting-edge data-driven methods. Yousef has published extensively on software requirement engineering, agile, and scrum. His recent publications focus on utilizing machine learning to predict work in progress for both newly formed teams with limited data as well as established teams.

Author 3: Dr. Mohammad Moshirpour

Dr. Mohammad Moshirpour is the director of the Software Engineering Practice and Education (SEPE) Research Group. He is an associate professor at the Department of Informatics, Donald Bren School of Information & Computer Sciences at the University of California, Irvine. His research interests include applied machine learning and AI, software design and development, requirements engineering, and software engineering training and education. He is the recipient of several awards for his work in developing tools and methodologies for software engineering practices and education including the 2021 D2L Teaching Innovation and the 2023 APEGA Excellence in Education awards. Dr. Moshirpour is also an adjunct associate professor at the Department of Electrical and Software Engineering at the University of Calgary.

Foreword

Post-secondary education has come a long way since the days of sitting in a lecture theatre for hours on end. Gone are the days of frantically writing down notes, class after class, hoping that you will be able to retain enough of that information so you can one day somehow use it all on the job.

We, as a society, have come to realize that the best way to educate and train the next generation, no matter the field of study, is to give them practical, hands-on tools and experiences. Not only will they be ready for their first day on the job, but they will have likely already seen what that first day will look like.

As engineering educators, it's our job to get students ready for industry. We need to give them the fundamental knowledge to understand the big picture of what they are working on, but also the nuances of the technology they are utilizing and the tools they need to get the job done. From there, they need to know how to transfer their applied knowledge and figure out where the pitfalls might be. Then it's time to get their hands dirty and put their knowledge to the test.

We want them to pursue greatness. We want them to succeed, but we also believe that lessons are to be learned in failure as well.

The journey from an idea and hypothesis to a finished product and conclusion is never linear, and no two journeys are ever the same. However, it is important for everyone to see how those stories unfold.

I've known Dr. Mohammad Moshirpour for a long time, as he was once a student at the Schulich School of Engineering, and I was his supervisor for his postdoctoral fellowship. From obtaining his Bachelor of Science in Software Engineering and Computer Science, all the way to receiving his PhD in 2016, before becoming a member of our faculty, he has been an influential person in our engineering and education community.

Mohammad is a leader in software engineering education and has gained national recognition for his work in curriculum and pedagogy design and development. He has won numerous awards, including the 2021 D2L Innovation Award from The Society for Teaching and Learning in Higher Education (STLHE), and the 2023 APEGA Summit Award in for Excellence in Education from the Association of Professional Engineers and Geologists of Alberta.

Mohammad was instrumental in building the master's in software engineering program, a one-year fast-track degree for students looking to specialize in the field after obtaining their engineering degree. He's also been integral in allowing students to think "outside the box" to explore technology, immediately paying dividends with courses that are built on industry collaborations.

At the heart of what he does, Mohammad is passionate about collaboration – not just within the post-secondary world but outside. He has helped numerous students gain practical, hands-on experience with local and international technology companies, which has contributed greatly to their career success.

Even after he moved on to the University of California, Irvine, Mohammad and I stay in regular contact. When he asked if I would write the foreword for his book, I was humbled and honoured because I have seen his own story play out right in front of my eyes.

In my opinion, there is no better person to help explain how a project can evolve from day one to completion. He knows about all the opportunities and growing pains, and how relationships are key to making today's students into tomorrow's change-makers.

Whether you're a new student or intern hoping to get a peek behind the curtains of the work needed to succeed in your industry, or you're an employer or leader who trains interns looking to get students and young people up and running faster, this book will show you how the right mix of foundational education, applied skill knowledge and soft skill development can create a lasting impression for everyone involved in your business or industry.

Albert Einstein once said, "Information is not knowledge. The only source of knowledge is experience. You need experience to gain wisdom."

I hope this book inspires you to provide an experience of a lifetime for the next generation of engineers.

Dr. William D. Rosehart, PhD, P. Eng., FEIC, FCSSE, FCAE
Professor and Dean
Schulich School of Engineering
University of Calgary

Acknowledgments

This book is the result of the collaboration between a community of practitioners, academics, and students and would not have been possible without the collective efforts of everyone involved.

Firstly, we would like to express our gratitude to Dr. Arash Afshar who played a key role as the industry advisor to guide our students in their learning journey. This work would not have been possible without his knowledge and generous guidance. We are very thankful to the students who collaborated with us on the Ignite project and greatly contributed to the development of this book. We are privileged to be included in their journey of discovery, from which we learned a great deal.

We are deeply grateful to our colleagues from academia (Dr. Behrouz Far, Dr. Laleh Behjat, and Dr. Kim Johnston), industry (Mike Bauer, Andy Czerwonka, and Beenish Khurshid) as well as our students (Richard Lee, Shamez Meghji, Rajpreet Gill, and Robert Brown) who reviewed this book and provided valuable feedback.

We would like to extend our appreciation to the Schulich School of Engineering at the University of Calgary, and Dean William Rosehart for their extensive support of the Software Engineering Education and Practice (SEPE) Research Group which made this book possible.

We would like to express our deepest appreciation to Wayne Sim, Quorum Software, and Peloton Well Focused for their generous contributions to SEPE and pedagogical research and development at the University of Calgary which partly supported this book.

Finally, we would like to thank our publisher for bringing this book to fruition. Their expertise, professionalism, and support have been instrumental in making this book a reality.

To Nahid and Ehsan, for your unwavering support and encouragement.

-   Niyousha

To my constant companion, Roya, who always believes in me, even when I don't believe in myself.

-   Yousef

To my lovely wife Mahsa and adorable daughter Hannah.

-   Mohammad

How to Read

The chapters are spread across different sections which directly correspond to the timeline of this internship project; The book starts off by introducing the Product Vision and Kickoff meeting with the Agile team and moves on to showcase their learning, growth, and efficiency throughout Training Week and development of the minimum viable product (MVP), to finally reflecting on lessons learned by the interns, educators, and industry professionals involved Post MVP.

The story is mainly told from the perspective of the first author who was an undergraduate software engineering (SE) student looking to apply for an internship at the time of the Ignite project's inception and development. To provide further context, most chapters contain the following elements:

A WORD FROM THE MENTOR

This section contains notes from the project leads, otherwise known as mentors in the Ignite project. In a nutshell, it allows for the mentors to elaborate on a particular concept or process as it is being covered in the story to provide further context for readers. The range of topics discussed in these sections include but are not limited to Agile terminology, Git policies, and general practices adopted by the Ignite team.

In the first couple of chapters, this section is used to introduce the Ignite team members i.e., developer interns, project leads, and product owners. As the story develops, this section is used to present the self-written reflections of the members throughout different phases of the project, from training week to development. These perspectives provided by students, educators, and industry professionals are paramount for gaining insight into some of the challenges encountered and lessons learned in an industry-academia cross-project.

This section is a textual representation of content being shown on a computer screen during team meetings, which is mainly comprised of issues created on GitHub.

Contents

Section I. Product Vision

1 Ignite 3

2 Conceptualization Through Feedback 10

Section II. Kickoff

3 Release Planning 19

4 Stories 29

5 Sprints 44

6 Stories to Tasks 53

7 Learning Curve 62

Section III. Training Week

8 Standups 73

9 Practices and Architecture 85

10 Grooming 96

11 Feature vs. Component Teams 108

Section IV. Spikes, Practises, and Guidelines

12 Continuous Integration 119

13 Spikes 128

14 Personas 139

Section V. MVP

15 Intern Support 151

16 Dependencies and Productivity 159

17 Refactoring 168

18 Overcommitment 180

19 Effective Communication 190

20 Striving for Agile Practices 197

21 Backends, Frontends, and Merge Conflicts 209

22 Design Debates 221

Section VI. Post-MVP

23 Deployment 237

24 Web Design 245

25 The Stakeholders 258

Epilogue 265

Section I. Product Vision

Chapter 1: Ignite

Monday, April 20

"The secret to getting ahead is getting started."

- Mark Twain

 Niyousha is an undergraduate student pursuing a Bachelor's in Software Engineering and Minor in Management and Society at the University of Calgary. Having just started her 3rd year, she is eager to learn about the many aspects of software development and has ventured outside of the classroom to pursue independent and collaborative research within the Department of Electrical and Software Engineering. Her current research interests include software engineering best practises, training, and design.

"There is another project I want to discuss with you."

I look up from my notes to my Surface screen where Mohammad's video feed shows him sipping from a cup of coffee. In the meantime, I try to take advantage of this pause to digest the prospect of taking on another summer project.

Given the recent global pandemic that's caused our university to halt all campus-related activities, it has been a hectic couple of weeks adjusting to doing everything from home. I had just finished writing the last of my online exams yesterday, finally marking the end of the Winter semester, and have already kickstarted my project for PURE, the Program for Undergraduate Research Experience funded by the University. It's a program designed to allow students to work with on-campus experts to learn how research projects are developed as well as how their results can contribute to new knowledge while solving relevant problems.

© The Author(s), under exclusive license to Springer Nature Singapore Pte Ltd. 2023
N. Raeesinejad et al., *The Ignite Project*,
https://doi.org/10.1007/978-981-19-4804-6_1

My research supervisor for the next sixteen weeks, Dr. Mohammad Moshirpour, has also been my professor for each software engineering course that I have completed since the beginning of my undergraduate program. He currently has a 4.5/5 rating on ratemyprofessors and most engineering students describe him as inspirational and respected in his field.[1] The reason for not saying *software engineering* students instead is because he is also one of the first instructors for all engineering students at the University of Calgary before they even decide which discipline to pursue. In fact, my decision to go into software engineering was heavily influenced after completing his introductory programming course during my first year, titled "Computing for Engineers" (ENGG 233). So, you could say the chance to embark on another research project with him is a pretty big deal for me.

 Mohammad is a professor in the Department of Electrical and Software Engineering at the University of Calgary. He is the director of the Software Engineering Practice and Education (SEPE) Research Group. His research interests include applied machine learning and AI, software design and development, requirements engineering, and software engineering training and education. He is a senior member of IEEE and the IEEE Chair of the Computer Chapter of the Southern Alberta Section.

"You've heard of Ignite, right?" Mohammad asks.

I nod instantly. Schulich Ignite is a club at the Schulich School of Engineering, where its members – mainly software and electrical engineering students - volunteer as mentors to teach basic programming to high school students using Python. They hold eight-week workshops each semester in the Schulich building after school hours. I had once visited one of the workshop sessions myself, courtesy of being invited by one of the mentors who was also a co-worker of mine. My first impression was that the students were very bright and eager to learn during the session, making me wish I could have also been acquainted with programming and more tech-related concepts earlier in my high school years.

"As you may know, I have been the director of the Schulich Ignite program since 2016. When we were organizing our Ignite sessions, we were really excited to see how the students liked them and their parents' reactions. We even have a gala at the end where parents come to see demos of their children's projects. So, we took this and asked

---

[1] ratemyprofessors is a review site which allows college and university students to assign ratings to professors and campuses of American, Canadian, and United Kingdom institutions.

ourselves, how could we scale this? What's the best way to scale this and reach more people in the community? It's not feasible to get University students to go out in the community themselves to teach more high school students.

"I also recently visited a high school with another professor to deliver a programming workshop to the students there. However, we ended up wasting a lot of time just installing the software and getting past all the firewalls on their computers. Many of the students were as disappointed as we were in the process. The logistics of this approach for visiting high schools all over the city is clearly not sustainable."

I frown. Surely there must be a better way to streamline their teaching approach. Perhaps they could deliver the workshops through some online video conferencing platform. Then again, this wouldn't really solve the issue of students having to spend extra time installing all the necessary software and downloading files for lectures. In fact, I wouldn't be surprised if more time would be wasted as a result of not having the professor or mentors around to guide them through the initial setup.

Echoing my thoughts, Mohammad reveals that the only solution is remote learning but with a twist. "What if there was a system that could provide both the instructor and the IDE on the browser?"

"This would promote both accessibility and convenience." I realize out loud.

Mohammad grins. "Exactly! Instead of me, as a professor, visiting a high school every semester to give a lecture on programming, the vision is that I would just be able to send a link to different high schools and everyone can enter my virtual classroom, inside which they may begin coding immediately. They'll see me on one side of their screen and their code on the other side. There is zero setup and zero overhead."

Sitting back in his chair, he continues. "We've been planning for a year now to build a website for delivering the workshops to the students online, to be able to scale the Ignite program and extend our reach far beyond Northwest Calgary. We were holding Ignite on campus which meant that attending these workshops was only convenient for residents in that quadrant."

I nod along passively, mentally entertaining myself with the prospect of several creative and logical aspects I could potentially be involved with in the design and development of the website. If that would be the case, this project is straight up my alley.

"The mandate of Ignite has always been diversity!"

The slight transition in Mohammad's tone from laid-back storytelling to what I recognize as his "lecture voice" stops me mid-nod, and I hurriedly flip through my notebook for a blank page to start jotting down the project's premises.

"This means we are teaching programming to those who may not ordinarily pursue careers in science and technology, thus *igniting* their interest. We achieve that through using technology, and online learning is the best approach to achieve this in a systematic, effective, and sustainable way. And what better way, than building a website as a team of software developers!

"I've got about a dozen teams of MEng students who are working with different companies for one of the software engineering bootcamp courses I teach."

Amid my frantic scribbling, I'm faintly reminded of an earlier conversation with Mohammad during one of our weekly research update meetings, where he had made a recommendation to read one of his published papers on "Designing a Programming Bootcamp for Non-Software Engineers". It was based on the new one-year Master of Engineering (MEng) in Software Engineering program launched by Schulich two years ago. I remember that integrated into the program is a one-semester "Bootcamp" semester - collaboratively designed by Mohammad alongside other instructors in the department - to ensure senior engineering students develop a strong foundation in programming and data analytics.

"You can almost think of this bootcamp as an MBA for Tech." Mohammad chuckles. "Anyway, I've arranged that a couple of these students will be working on building the Ignite website during these Spring and Summer semesters, which I think will be a great opportunity for applying their learning to a big-scale software project."

I try to maintain a decent level of eye contact with Mohammad's virtual visage as my pencil never once breaks contact with my notebook.

Mohammad continues without any breaks. "At the same time, my goal is to use this site-building process to bring practicality to research. We can take advantage of this internship-like environment by setting up metrics for some really cool data collection, like proof of training developers and tracking their progress."

As I scramble to finish noting down the last thought, he suddenly stops, and I glance up to check if his screen froze. It's happened a couple of times during our previous meetings, but we never cracked the case of whose internet provider was the culprit. What I see instead is him blinking at me expectantly. That's my queue.

"I think this is really interesting," I scramble for words, glancing to the side and trying to wrap my head around my own chicken scratch, "and I want to express my appreciation for allowing me to participate in this project as well."

"Of course." he smiles warmly.

"I just have one question." I continue with newfound expectance, "What kind of role did you have in mind for me?"

While I think I have a good idea of what the answer will be, I want to hear it from him directly. Already aware of Mohammad's many endeavors in collaborating with different teams of students, graduates, and industry professionals, it came to no surprise when he had revealed to me of having taken multiple PURE students alongside myself under his wing for the Summer. But judging from our past discussions, I know my role would be different from those of his other students. Thinking back to last September when we had our first formal meeting in his office, I remember how I had expressed my interest in publishing a paper in lieu of coding.

"You would be the first software engineering student on my team to actually like writing!"

I laughed along at the time but spent the next couple of months worrying over whether what I had said would work for or against my learning experience this Summer. At the time, having completed a couple of fundamental second-year software and computer engineering courses, it was natural to believe that my lack of knowledge and experience in software development would not be a good asset to offer when asking for an undergraduate research position on Mohammad's team. With my conviction that there will be many opportunities to improve my programming skills in upcoming courses, clubs, and internships through the remaining years of my academic career, I firmed my own resolve to leverage an alternative skillset.

Now, after having just completed the second year of my degree, I still share that resolve, but with the new-found knowledge that I would still gain valuable insight into the software industry. After all, you can't write papers about a topic in which you don't conduct ample research! Similarly, I'm intrigued to find out how much industry insight I could acquire from my role in this new project.

Mohammad takes another sip from his crystal teacup. "What I have in mind is for you to have a more central role, which involves monitoring this ecosystem and documenting the entire process on which the system will be built. Then, once we have collected all this data, we can generate a few good papers out of it."

"What kind of data are you aiming to collect?"

"Mostly daily blogs and reflections from the team, meeting notes and recordings, git commits and communication artifacts from Slack. There's a lot of subjects we can touch on for future papers because it won't just be about coding anymore, but also motivation, teamwork, efficiency of developer teams..."

I hum in understanding. Seems like I would just be a third-party observer after all. Despite my implied lack of participation in the actual building aspect of the website, I still find myself eager to learn and witness the process in its entirety. I mentally console myself that on the bright side, I can still become more familiarized with the common software development practices used in industry.

"By the way," I cut in before Mohammad continues to passionately list potential paper abstracts, "How many teams are there and how are they set up?"

"Right," he coughs, "So far, I'm thinking we'll have three agile teams, which will be mostly comprised of my MEng students and a couple of PURE students."

Noticing my surprised reaction, he says, "You know Niyousha, when people sign up for projects of big magnitude, there must be trust between them."

He waits for me to nod before continuing. "This is why I allowed the MEng students to form their own agile teams for the team design project course. These groups will be working part-time. Similarly, the undergraduate PURE students will form another agile team, however, they will be working full-time instead. Due to the innovative aspect of the Ignite system as well as the nature of PURE, the responsibilities of the undergraduate students will be heavily centered around conducting most of the research and development tasks, more so than just academic research. These tasks involve coming up with the most effective technologies, gathering data, and making decisions for the best way to employ some of the more innovative features of the website."

It all sounds like a win-win situation to me; on one hand, Mohammad has a rich team of engineering students of various backgrounds with good problem-solving and self-learning skills to bring forth his vision to engage high school students in an innovative way. On the other hand, the MEng and PURE students are provided with an ideal opportunity to not only apply their software engineering knowledge in a large-scale system with innovative features, but also a safe-to-fail setting to encourage research and creativity, as well as levitate some of the pressure that would have otherwise hindered their learning in an actual industry project.

"I have gathered two mentors for the team, and we already have a consulting company, Aranite, for maintaining the codebase and handling software development. Masoud is a graduate student who is acting as a project lead and Arash is the CEO of

Aranite. In addition, I have hired a seasoned industry professional, Yousef, who has tremendous experience in software process and management."

He concludes bashfully, "Including you, this is the dream team I had in mind."

I smile, trying to mask an emergent but familiar feeling of apprehension starting to cloud my excitement. Now that most of the details have been made clear to me, I can't help but already feel slightly overwhelmed. On one hand, just by hearing about the rest of these team members, I feel any dignifying amount of knowledge and experience I had before has now been truly rendered subpar in comparison. On the other hand, I remind myself that my own inadequacy makes for a steeper learning curve to come. I focus on setting my goal on the latter fact to appease my state of mind and accept the forthcoming challenge. All that's left is to meet the team and do everything in my power to keep up.

"This all sounds great and I'm looking forward to learning from the process!" I grin. "When do we start?"

Chapter 2: Conceptualization through Feedback

Tuesday, April 27

"No man who thinks ill, will hear the truth despite a hundred signs"

- Eblis

The week following my last discussion with Mohammad was not too eventful, other than the onslaught of emails to which I'm still getting the hang of replying in a timely manner. I have been invited to attend the first general meeting with the entire Ignite team which is supposed to be held this week on Friday. I'm assuming this means that all the mentors will be participating as well. To my dismay, I have been notified that the meeting will last for a minimum of three hours, every minute of which I will only observe and not participate in any discussion. Although unnerving, I try to not let this dim my original ambition to learn as much as I can from the process.

I recently attended another meeting last Friday with Mohammad and three academic professionals to further discuss the features of the new Ignite system. There were a lot of speculations and inquiries running through my mind when I received the meeting invitation from Mohammad a couple of days ago:

-----Original Appointment-----

From: Mohammad Moshirpour
To: Kat Dornian; Robyn Mae Paul; Niyousha Raeesinejad
Cc: Laleh Behjat
Sent: April 23, 9:53 AM
Subject: Ignite System Analytics
When: April 27, 8:30 AM-9:30 PM (UTC-07:00) Mountain Time (US & Canada).
Where: Zoom

© The Author(s), under exclusive license to Springer Nature Singapore Pte Ltd. 2023
N. Raeesinejad et al., *The Ignite Project*,
https://doi.org/10.1007/978-981-19-4804-6_2

Dear all,

As I have mentioned previously, we are currently building an online teaching platform for teaching programming. This platform will be used for outreach in Schulich Ignite and in teaching programming in various courses. The main points of strengths of this system include:

- Eliminating the need to install tools at the user end (since they will program right on the browser)
- Incorporating adaptive learning through providing just-in-time support for the students
- Incorporating active learning as students will be able to code along with the instructor (as they watch the live lecture) on the browser
- And of course, accessibility to our virtual classroom regardless of the users' location

An important capability that I hope to add to this system is data recording, data analytics and visualization to help us understand the efficacy of the tool and the teaching practices.

Here are the questions/challenges in this regard:

What data to record? Examples of this could include:

- Recording time taken for students to solve a challenge.
- Record the number of times a student asks questions during a session.
- Giving a short survey at the end of each session.

How to effectively maintain/increase student engagement (i.e., retention rate)? For example:

- How can we use gamification to spark user interest, such as developing a point system, etc.?
- How can we measure student interest?
- What sort of ethics approvals do we need as we provide these services to both our own students and high school students?
- How to learn from and visualize this data?

I think there is a tremendous opportunity here. I was wondering if you can please share your thoughts. Can you think of any other questions/challenges? Since we are developing the system as we speak, it would be great to have these points addressed. I look forward to our discussion on Monday.

Thank you,
Mohammad

Fortunately, I had already been acquainted with two of the invitees; Robyn is a Programming Planning and Evaluation Specialist at Schulich and Laleh is a professor in the Department of Electrical and Software Engineering with her many research interests including engineering education. Prior to the meeting, I had briefly researched the last unknown invitee and discovered that in addition to her TA position at Schulich, Kat is also Champion of Engineering Communication at TELUS Spark, the Calgary Science Center. According to an article on the University News homepage, she has played an active role collaborating with education research professionals in the past to organize computer science exhibits and public hackathons for K-12 children. Based on my impression from both past conversations and research, I had gathered that Mohammad's meeting guests are not only experts in their respective fields, but also advocates for advancements towards the new frontier of engineering education.

After reading the remaining body of the email within the context of this newfound knowledge, it seemed at the time that all the pieces had suddenly come together. I came to the realization that the scope of this project goes beyond what I had initially perceived, no longer bound by a couple of students programming a website but instead a business model, conceived through Mohammad's vision and kickstarted through active involvement of external stakeholders and experts to generate business value.

I was reminded of one of my past University courses in entrepreneurship where I learned how to identify and overcome confirmation bias throughout the design phase of new product development. Oftentimes, an entrepreneur, or really any individual with a new idea, has a certain tendency to seek things that confirm this idea, by starting off with strong assumptions and only engaging in talk with people who confirm them. This act of neglecting opportunities for open discussions results in merely building upon one's own beliefs and never encountering opposing ideas. As such, the individual becomes overconfident in their product's success and blind to evidence that would suggest otherwise, which is detrimental to new product development.

I had learned that the most effective way to overcome confirmation bias would be to perform the opposite act, by building a diverse community whose honest feedback will be tested against base assumptions, in order to test multiple alternatives to the new idea. It is important to note however that members of this community, typically comprised of advisors, potential suppliers, and other stakeholders, only provide tiny pieces of evidence, thus negating the chance of ever possessing a perfect report stating a new idea's exact success rate.

Bearing this notion of "advisors" in mind, I went into the meeting on Monday with the hope of becoming more familiar with innovative methods for gamification and data analytics which may be eventually applied in the Ignite system. Indeed, as I had expected, our discussion had yielded fruitful results. First and foremost, I became better acquainted with two particular objectives of the Ignite platform towards providing an optimal learning experience for students: The first objective is to increase student engagement and lesson retention rate and encourage thrive by effectively using gamification and rewarding students for their achievements; The second objective is to enable the analysis of the efficacy of teaching approaches and student engagement through smart data recording, data analytics capabilities, and data visualization.

Laleh, Robyn, and Kat suggested rather innovative ideas to achieve these objectives, the most prominent of which, in my perspective, was the idea of a literal "concept map" for students to visually assess their advancements through each milestone i.e., learning outcome in their course. This feature may be extended to further courses in the long run, once the Ignite platform is proven to be dependable for supporting programming lessons. Then, the concept map could be implemented as a template for any instructor to be able to customize on the platform, with their own course-related design.

Certain broader ideas were also brought up, such as the prospect of introducing a social network aspect towards rewarding student progress. Another involved hosting both internal and external competitions like virtual hackathons through the platform to further engage students in problem-solving, working in teams, and applying the knowledge and skills they have learned in sessions towards developing practical solutions. Introducing a live point system to these friendly competitions or even regular activities during their lessons may also increase student engagement and retention.

After having sat through the meeting and heard all these ideas to enhance the teaching and learning experiences of students on the Ignite platform, I'm developing a deeper appreciation for domain/subject matter experts in a project of this nature. Regardless of their indirect involvement in the development of the system, the implications of their knowledge and expertise can be of great value for the agile teams to make better informed decisions regarding the high-level features of such system. Initially, I had expected that the product owner, that is, the person who first came up with the business idea would – which in this case is Mohammad – would possess the most knowledge of not only the business process, but also the nitty gritty details of the system. However, certain areas of behaviour can be further explored and solidified with the help of domain experts. This is where feedback can be most useful towards system conceptualization.

Another recurring thing that I noticed during the meeting was that the domain experts also noted many opportunities for collecting data from student and instructor interactions with the system for every one of these suggestions they had made. From my understanding, Mohammad's intention is to not only build the Ignite platform's features to uphold functionality requirements, but also in ways that would allow for him to capitalize on the potential data they can bring in. Regarding the gamification and data analysis features suggested in our meeting, I wonder if Mohammad plans to integrate some or all of them in the Ignite system, or to modify the definitions of certain features that have already been planned. If so, I would assume that they would be presented to the team prior to the general meeting next week as well. Whether that is the case or not, I am left wondering: What is the extent of the development team's involvement in the planning process for the features of the product being developed?

Mohammad

The main objective of the Ignite system is to create an environment where we can simulate a live classroom using a mentorship structure. Students first come into my live lecture where they can code alongside me, then get divided into breakout group to work through exercises, discuss with their peers, and ask for help from the mentors at any time along the way if they get stuck on a problem. For now, I'm considering the ratio of mentors to mentees as one to five. This way, the students can learn programming fast in a very efficient way through promoting the core elements of Ignite: active, adaptive, and project-based learning!

Developing the Ignite system is also a great opportunity for my students to advance on their paths to becoming full-fledged software engineers. My MEng students in particular will be working part-time on developing the system as part of their final project for the courses ENSF 609 and 610, otherwise known as Team Design Project in Software Engineering I and II, which will span over the coming Spring and Summer semesters. These two project courses will bring an end to their year-long MEng program. I am treating this set of courses more as a mini-internship, like a capstone instead of what it used to be, which was a mini-thesis with heavy emphasis on students reporting what they have done in the course. I had noticed that very few MEng projects in the past were in collaboration with industry. For this program, every team works with an industry partner like Aranite, with the exception of one team which are working by themselves on a startup project. Through these industry-lead projects, I am aiming to place heavier emphasis on the role of industry in academia with higher priority on efficiency rather than documentation.

Section II. Kickoff

Chapter 3: Release Planning

Friday, May 1

"To be uncertain is to be uncomfortable, but to be certain is to be ridiculous."

- Chinese proverb

1.00 p.m.

I paste the Zoom meeting URL in a new tab on my browser and mentally brace myself for a long afternoon while it loads. A new window pops up and prompts me to connect my audio. Under the impression that I will be the only new face for the team, I enable my video as well. Immediately, two other videos appear next to mine in a grid layout. The first belongs to Mohammad who is typing and staring into what seems to be another computer screen. The other video shows Yousef with a pensive gaze.

 Yousef is one of the mentors on the Ignite Team. He is a Ph.D. candidate in Software Engineering at the University of Calgary with experience as a trainer and coach in the Software development industry. His current research is on improving software development processes using machine learning.

He suddenly starts to laugh after noticing who I am and I can't help but reciprocate, overcoming my initial shock. It seems we didn't really need to have an e-introduction meeting after all; Yousef is in fact a mutual acquaintance who has offered guidance for my Software Engineering courses in the past.

"It's great that you two already know each other!" Mohammad chuckles along and we both nod in relief.

Before we can continue chatting, more participants enter the meeting. I quickly recognize the name of one of them whose video is enabled, showing a man in a hoodie and glasses.

Arash is the founder of Aranite and has over a decade of experience as a Software Engineering industry. He has a PhD in Computer Science from the University of Calgary. In both his industry and academic experience, Arash has mentored many junior developers. As a result, one of the key pillars of Aranite is training junior developers.

More casually dressed people show up on video, but I don't recognize them. Seeing as how nearly everyone except for Arash and Mohammad have their microphones muted, I follow their example. After a few seconds, the participant count at the bottom of my screen finally stops incrementing at nineteen.

"Looks like everyone's here," Arash announces, a low resonance to his voice. Suddenly, the grid layout containing everyone's videos is replaced with the view of a Chrome browser, on top of which there's a notification indicating that Arash is sharing his screen. His current tab, amongst many others, is open on GitHub[2]. Although minimal, I have some prior experience with this platform in one of my recently completed Software Engineering courses with Mohammad as my professor; While working on its end-of-term project, my partner and I frequently uploaded our code and other related files on this website. We even made use of one of the provided project board templates (I think it was called a Kanban board) to keep track of who did what. Because we were only briefly introduced to some of its basic features, I'm eager to discover more features and how they will be utilized for a project of larger scale, complexity, and teams.

Arash kicks off the meeting by navigating to a repository named 🖥ignite and scrolls down to the README.md file.[3]

"These are the contributing guidelines which I will be demonstrating in today's meeting. We can ignore the initial setup part for now."

Before I get a chance to read the descriptions of the guidelines, he immediately opens a folder in the repository called how-to which contains different markdown files like domain-model.md, how-we-work.md, python-guideline.md, react-guideline.md, and release-planning.md. He opens how-we-work.md and I quickly browse over its content as

---

[2] GitHub, Inc. provides hosting for software development and version control using Git. It offers the distributed version control and source code management functionality of Git, plus its own features. It provides access control and several collaboration features such as bug tracking, feature requests, task management, and wikis for every project.

[3] Git is a distributed version-control system for tracking changes in source code during software development. It is designed for coordinating work among programmers, but it can be used to track changes in any set of files.

Repositories in Git contain a collection of files of various versions of a project. These files are imported from the repository into the local server of the user for further updates and modifications in the content of the file. A Version Control System (VCS) is used to create these versions and store them in a specific place termed as a repository.

he scrolls down through headers labeled "Team Structure", "Roles", "Artifacts", "Activities", etc., but most of the terms I come across are unknown to me.

"As you may all remember from our first meeting, we talked about the goal of the Ignite project and some agile concepts. I asked you all to look into the rest of these how-to files and familiarize yourselves with most of the terms to come prepared to this meeting. Therefore, I will not be going through each term one-by-one today, but I will be using them, so if you have questions at any time, feel free to stop me and ask." I'm a little disappointed that he won't go over those unknown terms, but I'm assuming everyone except for me is already familiar with them and there must be much more material to cover in this meeting.

The next document that Arash opens is release-planning.md which he explains will be the main document that the team is going to work on for today.

A WORD FROM THE MENTORS

A **release** is a combination of features packaged together as a coherent deliverable for users. Releases represent the rhythm of business-value delivery and should align with defined business cycles.

Release planning, otherwise known as "long-term planning" or "medium-term planning", involves making a prediction of what would be delivered by certain time periods. When planning for releases, we answer questions such as:

> → "When will we be done?"
> → "Which features can I get by the end of the deadline?"
> → "How much will this cost?"

Release planning must balance customer value and overall quality against the constraints of scope, schedule, and budget. The output of release planning is a **release plan**.

"Before we start, does anyone have any questions about the terminologies, the process, or anything you have learned or heard so far?"

A few seconds of silence pass while I wallow in my lack of knowledge on all of them.

"Ok, I'm hearing none." Arash switches his tab to a PowerPoint presentation. As he starts scrolling through the pages, a deep but hesitant voice breaks through.

"Umm, I have a question." Toya announces.

 Toya is one of the Software Engineering graduates in the MEng program at the University of Calgary. He recently graduated in Mechanical Engineering at the University of Alberta in 2018 and was part of the first MEng Software Engineering cohort before taking a leave of absence to work at Suncor as a co-op student. His main interests are frontend development, robotics and AR/VR.

After guiding Arash back to the tab with the release planning.md document, he asks "In the stories section, course management wasn't initially part of the stories that you have in there, so is it something that's coming down later?"

"Yes," Arash confirms, "What we have here is our MVP for June 15th. After that, we will work on the enhancements which course management is a part of." [4]

Toya thanks him and Arash waits for any further questions while switching back to the PowerPoint tab. After another moment of silence, he continues, "That was actually a good segway into the beginning of this presentation." He proceeds to explain the contents of the first slide, titled "Key Concepts of Managing a Project".

A WORD FROM THE MENTORS

In software development, it is important to distinguish between the following terms:

→ A **milestone** is a significant point in a project that must be well-defined.
→ A **feature** is a service that fulfills a stakeholder's need.
→ A **product backlog** is a list of features, defects, or technical work that is valuable from the product owner's perspective. It is the single authoritative source for things that the development team works on. That means that nothing gets done that isn't on the product backlog!
→ A **story** is a short description of a small piece of desired functionality, written in a way that is easily understood by the user.

"We have a concept of milestones and releases, but we use these words interchangeably. For example, the MVP is our first milestone, the first release. Each milestone or release will have a set of features. We generally start with a very high-level set of features. For example, we want to have video streaming in a course, and this is just a one-line description. But then, during release planning, we would describe it in more detail and break it down into smaller stories which we can then talk about in terms of cost, time,

---

[4] A Minimum Viable Product (MVP) is a version of a new product that allows the team to collect the maximum amount of validated learning about customers with the least effort. This validated learning comes in the form of whether your customers will actually purchase your product. Eric Ries.

complexity, dependencies, or whether it's suitable for this release. These are the things we usually talk about when planning a release."

He announces that this meeting is dedicated to the full planning of the team's first release, where they will briefly go over what he has planned. This way, the team won't be overloaded with information from the very beginning.

"These high-level headings that you see here – like *lesson creation and user management* - are features that we want to be completed for the MVP." He continues to highlight more headings:

- Video streaming
- Raise hand and ask question
- Chat with TA and instructor
- Edit the code of individuals by TA

"As Toya mentioned, features such as course management, data analytics visualizations, or a lot of the editing ones are left out of this feature set. That is because we want to have something that works and is deployable, which we can then call a product. This is what we are aiming for in the first release." He scrolls back up to the first mentioned feature:

Lesson creation and user management
User management

"Once we have these sets of features, then during the planning sessions, we will try to break them into smaller stories. For example, one of the user management stories is for authentication, where we want students and the instructor to login using their email or other social media platforms. This is what we often call social login, or OAuth." [5]

Authentication (social login):

"Now this on its own is not very descriptive. If I tell someone to go implement authentication using Google, it's not very clear what needs to be done. That is why we break down this short description into more details. The first thing we need to have is what the feature is supposed to look like from the user's perspective."

---

[5] OAuth is an open standard for access delegation, commonly used as a way for Internet users to grant websites or applications access to their information on other websites but without giving them the passwords. Generally, OAuth provides clients a "secure delegated access" to server resources on behalf of a resource owner. Wikipedia.org

---

A WORD FROM THE MENTORS

A user story template provides us with a common format for writing user stories. The most recommended template is often referred to as **"As a… I want to… So That…"**, where:

→ **As a** (who wants to accomplish something)
→ **I want to** (what they want to accomplish)
→ **So that** (why they want to accomplish that thing)

We add **acceptance criteria (AC)** to define a desired behavior in detail and determine whether a user story has been successfully developed. These are our "conditions of satisfaction" or "exit criteria" that our platform must satisfy in order to be accepted by our users and other stakeholders!

As a user I want to login using my Google account so that I can access the workshop.

"In this sentence, we are explicitly calling out Google instead of just saying social login. The reason for doing this is so that the user can understand the business value of it. This is good because it is describing the purpose of the story, however it is still not enough; It still lacks definitions of more fine-grained criteria for this story. At this point, we define a set of acceptance criteria."

- AC1: If a user does not have an account, she can register using her google account
- AC2: If an instructor logs in using her google account, she should see the "instructor" role in her profile
- AC3: Once the user registers, an empty user profile is created
- AC4: As long as the user profile is empty, a warning banner should be shown so that the user can click and update it

"This is better since it provides more information. However, there is still room for improvement since the story can be broken down further. We call this level of granularity a story and the AC will be used for implementing and testing our functions. Now, the agile team will start from a story like this and split it into smaller tasks that fit their team structure and capabilities."

A WORD FROM THE MENTORS

A **task** is the technical work that a development team performs in order to complete a story. Most tasks are defined to be small. Team members usually define their own task-level work.

"For example, one member could do the registration part in AC1 and another could work on displaying the warning banner in AC4, so both the backend and UI sections can be developed in parallel. The other two AC cannot be done at this time, since they depend on AC1 and AC4. From this, 3 tasks can probably be created - one for the UI, one for the backend, and one for the user management in AC2 and AC3 - to be picked up and implemented by the team members after AC1 and AC4 are done."

Arash takes a pause before asking the team if they have any questions on the distinctions between the whole feature, the stories that build up that feature, and the tasks that build up a story. Following a short period of silence, he continues.

"We still need to be able to size these stories and tasks so that we have a sense of how hard they are and discuss whether we are thinking of the same thing or not. For stories, we will be using the Fibonacci points as a unit for measuring their complexity."

A WORD FROM THE MENTORS

Size refers to how much effort is needed to finish a given user story or task. Some teams estimate the size of user stories with a predefined set of values. One commonly used set is the modified version of the **Fibonacci sequence**:

SIZE	INTERPRETATION
1, 2, 3	Small
5, 8, 13*	Medium
20, 40	Large
100	Very large (often regarded as a feature)
∞	So large that it doesn't even make sense to put a number on it
?	Additional clarification is required to size the story

*For many teams, 13 is the maximum story size scheduled into one sprint. They would break down any story larger than 13 into a set of smaller stories with sizes consistent with the Fibonacci sequence. Each Fibonacci number may be expressed as the sum of its 2 previous numbers in the sequence (e.g., 13 = 8 + 5).

Why do we use the Fibonacci sequence? After the number "2", each number is about 60% larger than the one before it. According to Weber's Law, if we can distinguish a 60% difference

---

in effort between two estimates, we can distinguish that same percentage difference between other estimates.[6] Since each number increases by the same noticeable proportion each time, our team can distinguish estimates easier! [7]

"For the purpose of our team, sizes of 0, 1, and 2 refer to something trivial. 3 is easy, 5 is medium, and 8 is hard. Anything more than a size of 8 indicates that the story is too big and needs to be broken down into smaller stories.

"I have mentioned this in our last meeting, and I will mention it again. These points and hours are *not* commitments. If you say something will take 1 hour, and you spend 2 hours on it instead, you have not done something wrong! Estimations are almost always wrong. What we are trying to do here is make sure that everyone is on the same page and identify areas that need to be discussed further."

Switching back to the PowerPoint, Arash summarises what he has covered so far. He suddenly admits he forgot to mention the last point on his slide about when the planning sessions will occur.

"All of these things that I have explained happen at the beginning of a sprint."

A WORD FROM THE MENTORS

A **sprint** is a timeboxed iteration with a short duration. Typically, a timebox between one week and a calendar month during which the development team is focused on producing a potentially shippable product increment (increment in short) that meets the Scrum team's agreed-upon definition of done.

"Our first sprint starts on Monday. We have already done the planning and we have a set of user stories. We are going to talk about those stories and then size them, break them into tasks, and size those tasks. This is what we will do in today's planning session as well as all future sessions."

He pauses for a bit before noting, "When we say we want to work on a story during a sprint, it means that it should be finished by the end of that sprint. At which point, we should have deployable code. If something is half done, such as the backend being done

---

[6] Ernst Weber, a 19th century experimental psychologist, observed that the size of the difference threshold appeared to be lawfully related to initial stimulus magnitude. This relationship, currently known as Weber's Law, states that the difference we can identify between objects is given by a percentage.

[7] This is known as the "Difference Threshold" or "Just Noticeable Difference", which is the minimum amount by which stimulus intensity must be changed in order to produce a noticeable variation in sensory experience.

but the frontend not done, then that code, in its half-completed state, will not be merged and instead of receiving points, you will move that story to the next sprint. Thus, at the end of the sprint, whatever is in the repository should be deployable and runnable code."

A WORD FROM THE MENTORS

If a story is not done by the end of a sprint, it should first be moved back to the product backlog. If that story still has priority, it may be planned for the next sprint. Work never moves automatically from one sprint to the next!

If you are going to finish the story in the next sprint, just move it along and don't rewrite its description and AC. If the remaining work will be deferred to a later sprint, write a new story that describes just that subset of functionality.

The next slide contains an image of what seems to be the entire release planning process:

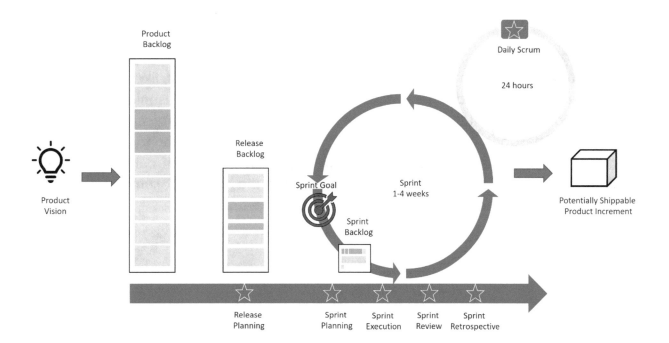

A WORD FROM THE MENTORS

During **sprint planning**, the development team gathers to agree on a sprint goal and determine what subset of the product backlog it can deliver during the forthcoming sprint.

One of the outputs of sprint planning is a **sprint backlog**, which is a list of product backlog items pulled into a sprint along with a plan for how to achieve them. This is frequently expressed in terms of tasks.

"From the release planning, we create a product backlog. In the sprint planning, we start from the top of that backlog, break it into tasks, and bring them into each sprint, which lasts one week for us. At the end of the sprint, we discuss and review our work. I will be covering how this is done in later slides."

Looking at the diagram, I'm surprised to discover even more terminologies with the word *sprint* included in them. I have never heard of it being used in this context before, and even after hearing Arash use it few times in his explanations, I still do not have a full understanding of why such a term exists. Apparently, a sprint represents a duration of one week for the Ignite team. In that case, why not just keep referring to it as a week? Hopefully, the reason will be cleared up for me soon.

For now, based on its connotation, my educated guess towards what sprint planning entails is trying to plan upcoming work for a team as fast as possible, like a runner trying to finish a sprint in record time. I could justify this definition by reasoning out that the ends are more important than the means in software development projects, and planning work should be done as quick as possible in a sprint-like manner to mitigate over-planning and promote tangible results. However, given my lack of confidence in this assumption, I make a mental note to myself to read all the how-to documents and brush up on these terminologies to be prepared for future meetings.

Chapter 4: Stories

Friday, May 1

"The engineers of the future will be poets."

- Terence McKenna

Arash proceeds to the next presentation slide, which covers his plan for the team today.

"I have already talked about the general features of our MVP. Now, we are going to simulate everything I have talked about, regarding how a story is created on GitHub, sized, implemented, and merged. To facilitate this simulation, I will now ask Yousef to go through an example with me."

After Yousef nods in agreement, he pulls up the <u>release-planning.md</u> document once again.

"Okay, we are going to do something very simple - something that doesn't require any technical knowledge - just so you can learn how the process works."

Process

Each team chooses a story and gives it story points.

- Split the story into tasks
- Estimate the story based on complexity, effort and uncertainty (non-committing)

"You will notice that the <u>release-planning.md</u> document has a *Process* section. However, Yousef has already created this section along with an in-depth explanation in another file called the <u>how-to.md</u> document. Since it doesn't make sense to have this small section duplicated here, we want to remove it. The feature that arises from this entails cleaning up the <u>release-planning.md</u> document. Let's start by creating a story for that."

N. Raeesinejad et al., *The Ignite Project*,
https://doi.org/10.1007/978-981-19-4804-6_4

He navigates back to his GitHub tab and opens a new issue: ⊙`Cleanup release-planning document`.

"The first thing we do is describe this in the point of view of the user."

> As the Product Owner, I want to clean up the release-planning document so that the document only contains the stories and vision.

"Do you agree with this description, Yousef?"

"Sure," Yousef agrees.

"Next, we need to list the acceptance criteria."

"They should be clear."

"Yes, so as you write more ACs down the road, please try to be more explicit."

> AC: Remove the section "Process" of release-planning.md

"That's good enough for the acceptance criteria. Now, we're going to add some bookkeeping information for this issue."

He proceeds to assign the "story" `Label` to the issue and then adds it to ▢`ignite` with "MVP" as the `Milestone`.

"Now we need to size it. For that, we can use this website." He opens a new <u>scrumpoker</u> tab on his browser.

A WORD FROM THE MENTORS

Planning Poker is a consensus-based and playful technique for the relative sizing of PBIs.

The team first meets in the presence of the customer or PO. Around the table, each team member holds a set of playing cards with numerical values appropriate for point estimation of a user story. The PO briefly states the intent and value of a story. Each member of the development team silently picks an estimate and readies the corresponding card, face down. When everyone has taken their pick, the cards are turned face up and the estimates are read aloud.

> The members who have given the highest and lowest estimates justify their reasoning. After a brief discussion, the team may seek convergence toward a consensus estimate by repeating this round as many times as necessary.
>
> There are several available tools for playing planning poker, such as physical card decks, websites, and applications!

"This online tool allows for everyone on the team to submit their own size estimates, which will only be revealed to the rest of the team once they are all submitted." To demonstrate his point, Arash and Yousef do a quick demo of choosing point cards for the issue via the same link on the website. Both of their cards reveal an estimate of 1.

"Once we have discussed any differences in the points and reached a consensus, we can assign the proper size to the issue."

Apply labels to this issue

point-

point-easy
scrum story point 3

point-hard
scrum story point 8

point-medium
scrum story point 5

point-spike
scrum story point - spike

point-trivial
scrum story point (0-2 range)

Create new label "point-"

Edit labels

"Now that we have our story, the development team gets together and tries to split it into tasks. Tasks and stories are both issues on GitHub and the only way they are differentiated is through this bookkeeping information, such as labels and sizes."

He then creates another issue: ⊙ `Remove process from release-planning`.

"Tasks don't follow any particular format or guideline, since they are created by the teams who will decide the format themselves. One requirement for each task is to include an estimate of how long it will take, which you can determine through scrum poker, regular chat, or any other method."

Arash then asks how long their example issue should take, to which Yousef jokingly responds a couple minutes. He chuckles and types it as so in the issue description.

Estimate: 10 minutes

"Could you explain why sizes would be different in general?" Yousef prompts Arash.

"Regarding both stories and tasks, we are not just discussing the problem, but also the solution. For example, in this case, I am saying that the solution is to simply remove a couple of lines. But Yousef may have a different solution in mind; Maybe, he is seeing some dependencies that I have overlooked or am unaware of. As a result, he may have a different estimate. These differences in thinking trigger discussions about which solutions are better."

He mentions that if the team already knows what the solution is, they can also include it in the brief task description after their estimate.

Remove process from release-planning

Estimate: 10 minutes
Going to delete some lines.

After assigning the proper labels to the issue and submitting it, he adds that for tracking purposes, to link the issue back to its parent - the main story - its link may also be included in the story description:

Cleanup the release-planning document

As the Product Owner, I want to clean up the release-planning document so that the document only contains the stories and vision.

AC:
- Remove the section "Process" of release-planning.md

Tasks:
- https://github.com/aranite-open/ignite/issues/25

"Now, other tools like Jira let you do all of this in the UI, allowing to assign subtasks to issues. We are just doing some manual work here within the limitations of GitHub." [8]

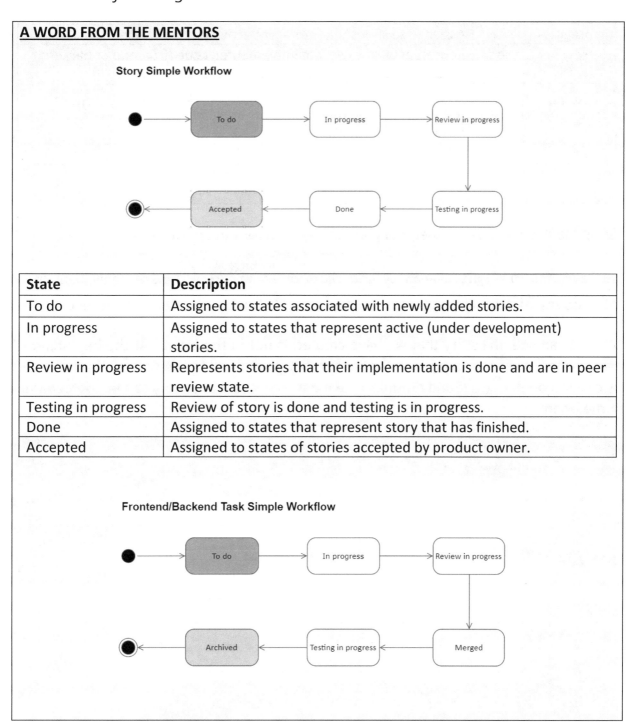

A WORD FROM THE MENTORS

Story Simple Workflow

State	Description
To do	Assigned to states associated with newly added stories.
In progress	Assigned to states that represent active (under development) stories.
Review in progress	Represents stories that their implementation is done and are in peer review state.
Testing in progress	Review of story is done and testing is in progress.
Done	Assigned to states that represent story that has finished.
Accepted	Assigned to states of stories accepted by product owner.

Frontend/Backend Task Simple Workflow

---

[8] Jira (/ˈdʒiːrə/ JEE-rə) is a proprietary issue tracking product developed by Atlassian that allows bug tracking and agile project management. The product name is a truncation of Gojira, the Japanese word for Godzilla. The name originated from a nickname that Atlassian developers used to refer to Bugzilla, which was previously used internally for bug-tracking.

State	Description
To do	Assigned to states associated with newly added tasks.
In progress	Assigned to states that represent active (under development) tasks.
Review in progress	Represents that review of implemented code (frontend or backend) is in progress.
Merged	The code of task was merged into story branch.
Testing in progress	Assigned to states that represents implemented backend or frontend is under test.
Archived	Assigned to states of stories accepted by product owner.

"Suppose that we have finished our planning session for this sprint. Let's go to our board."

He navigates to 🖥**ignite-aranite** and clicks on `Projects`. 🔲**Ignite** only contains 2 other issues: ⓘ**Adding jupyterhub docker** and ⓘ**Cleanup the code structure**.

"As you can see, the story that we have created is not in the board. At the beginning of every sprint, you need to bring in the stories that you added to the sprint." He shows this by clicking on `Add cards` and dragging the most recent task and story to the `To do` column in the board:

"Let's say I am a developer and I want to work on this task. What I need to do first is drag it to the `In progress` column and then assign it to myself. This is so that we know who's working on what and won't have two members working on the same thing." Opening the task, he clicks on *Assignees* and finds his own profile to add.

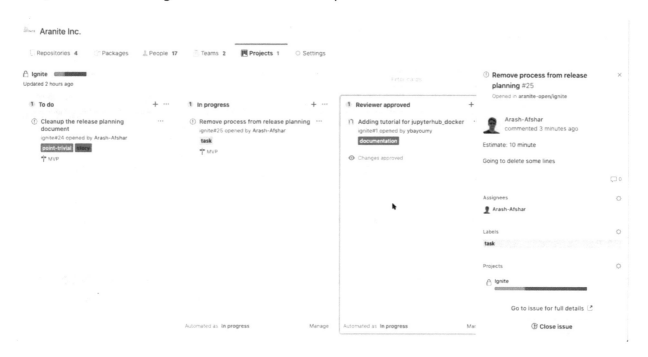

"I will demonstrate shortly why the issue numbers are important." He takes another pause for any questions from the team. This time, a question pops up in the Zoom chat.

> **Kevin:** I am a little confused about who is responsible for creating issues.

Kevin is a Developer on the Ignite Team. He had years of experience working at a web design firm and did extensive research analysis on UX/UI's impact. Currently, he is working towards his MEng degree in Software Engineering at the University of Calgary.

Arash reads the question out loud once for everyone before answering it.

"Firstly, there are two general types of issues: stories and tasks. Stories are usually created either by me or the development team during planning sessions. As you're working on something, you may come up with a new feature or you find a bug. All of these are issues you may create, which are effectively stories. The second type of issues are tasks which are created by the development teams themselves. At the end of a planning session or the beginning of a sprint, each team gets together and decides how they want to split a story into tasks, for which they will create issues."

"Hey, Arash?" Peter speaks up.

 Peter is one of the student developers on the Ignite team. He is currently finishing up his MEng in Software Engineering at University of Calgary, having prior experience in Alberta and Ontario as a mechanical engineer in a project engineering role.

"Who has permission to assign tasks? Can anybody assign tasks to anyone else?" Peter asks.

"In agile, nobody dictates who should work on what. The teams are self-organizing, meaning that from the set of stories for your team, you pick and assign one for yourself. I will try not to assign stories to other people."

A WORD FROM THE MENTORS

"The best architectures, requirements, and designs emerge from self-organizing teams."
- 11th principle of Agile

Self-organization is a property of a team that organizes itself over time, without an external dominating force. It reflects the management philosophy whereby operational decisions are delegated as much as possible to those who have the most detailed knowledge of the consequences and practicalities associated with those decisions. It is totally different from traditional top-down, command-and-control management.

The benefit of allowing a team to self-organize isn't that the team finds some optimal organization for their work that a manager may have missed. Rather, the team is encouraged to be fully responsible for performing their work.

Arash returns to the project board to create another issue, ⓘ **Remove the newlines after the process is removed**, to simulate a story having multiple tasks.

"By the way," he notes while typing an estimate of 5 minutes in the task description, "Most tasks will take a couple hours or days, so don't break them down into minute-level granularity."

He also explains that tasks can be defined according to what makes sense for their parent story, such as backend, frontend, or database portions.

He then initiates a demo on his terminal to show how to follow the instructions outlined in the *contributing guidelines* section of the README.md file.

Example Workflow:

1. Create an issue (assign proper labels and link to the parent issue
2. Bring your local master branch up to date with origin
3. Create branch "issue-NUM/descriptive-name" (e.g., issue-123/backend and issue-123/frontend, etc.)
4. In case of dependency, branch from another issue branch instead of master
5. Test the work by merging it back to issue-123/main and submit Pull Request
6. Each PR should be reviewed by at least one mentor
7. Definition of Done (DoD)
   - In the first few sprints: Simply submit PR following the above rules
   - Later on:
     - Add integration tests
     - Remove branches of finished stories
     - Etc.
8. IMPORTANT: Squash and merge (do not add merge commit)
9. Sprint review
   - Arash will pull master on his machine
   - In the Zoom call, each team will get control of the mouse and present their work on Arash's machine

A WORD FROM THE MENTORS

Visual process of GitHub flow for developing a story in the team.

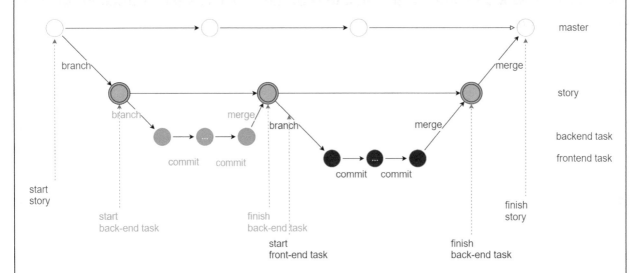

There are four branches in the image:
1) master
2) story branch: a branch for story created from master branch
3) backend branch: a branch for backend task created from story branch
4) frontend branch: a branch for frontend task created from story branch

"The first thing I need to do is to checkout master and bring my code up to date," he explains as he enters the corresponding commands in his terminal. "I will fetch from the master branch of the origin repository. Then, I need to reset my local master branch to what I have just fetched. This gets rid of everything in master, so don't do any development in master or push anything to it."

Returning to the project board, Arash refers to the issue number of the story he recently created.

"I can now work on this issue." He checks out a new branch with the specific issue number and creates another branch from it, reminding everyone once again that they should not work on the main branch, since two people could be working on its subtasks at the same time. He proceeds to quickly finish the task he created earlier by removing a couple of lines from the release-planning.md document and commits his changes.

"These changes would not be visible to someone working on another subtask for the same story and having not fetched my work yet," he notes.

Finally, he shows how each pushed branch may be submitted as a Pull Request (PR) on GitHub.

"However, as I have written in the contributing guidelines, we don't want to push unfinished changes," he reminds. "We want to submit PRs only for completed features. So, first merge your changes to your main story branch, then submit one PR for all of them."

However, he is quick to point out that this is not a hard rule. "This is just to make sure that in the beginning, you get a hang of how your work can affect that of other members. It may seem excessive in the beginning with all the counterintuitive steps like too much branching, but as we progress and as the teams mature, in around 2-3 weeks' time, these PRs may be submitted separately without risk of too many merge conflicts."

A WORD FROM THE MENTORS

Version control systems like Git are all about managing contributions between multiple distributed developers. Sometimes multiple developers may try to edit the same content.

For example, if Developer A tries to edit code that Developer B is editing, a conflict may occur. To alleviate this, developers work in separate isolated branches, but when they try to merge their own code, a conflict may still occur.

"Now that both features are up, one of the members will take on the role of consolidating both of these features into one branch and submitting it in a PR." He merges all his changes, adding on to his reminder that this allows for team members to not step on each other's toes in testing. He assures that if this restriction causes more conflict, they can adopt a more relaxed approach towards submitting PRs instead.

He then shows how to link a PR to an issue and submit it on GitHub. After assigning Yousef as a reviewer to review his PR, he tells the team to not worry about the other labels. [9]

---

[9] In software development, peer review is a type of software review in which a work product (e.g. document, code, etc.) is examined by its author and at least one more colleague to evaluate its technical content and quality. According to the Capability Maturity Model, the purpose of a peer review is to provide "a disciplined engineering practice for detecting and correcting defects in software artifacts and preventing their leakage into field operations".

Next, he pulls up the project board to show how the issues are now linked by one PR:

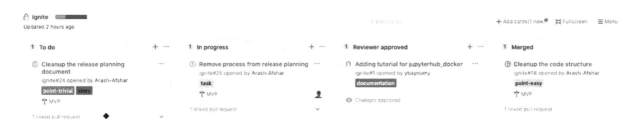

Arash requests for Yousef to share his screen next to show how the approval process works on the reviewer's side. Yousef proceeds to quickly demonstrate this by clicking on *Files Changes* and submitting his approval of the PR. Arash thanks him and continues with his explanation.

"After you submit your pull request and it is approved, make sure to move it to the `Reviewer Approved` column on the project board so that others know you are waiting for a review."

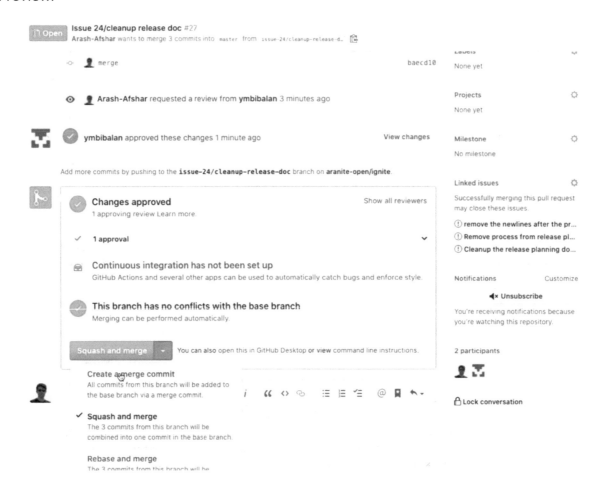

"One very important note about merging," Arash mentions, "By default, it is *in create a merge commit* mode. We don't want that. For the first time, please click *on Squash and merge* and this will become the new default from that point.[10] The reason for this is that I don't really want these three commits since they don't add any value in terms of the Git history."

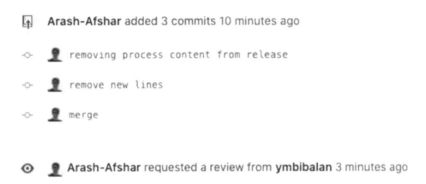

"As you work more, you will see that there will be several commits associated with one PR, such as moving something, fixing a bug, commenting during a review, and so on. Over time, this will develop into one big, ugly commit history. If you use *Squash and merge*, all you're caring about is completing the task."

"Notice now that all of these issues are in Merged."

---

[10] When you select the *Squash and merge* option on a pull request on GitHub, the pull request's commits are squashed into a single commit. Instead of seeing all individual commits of a contributor from a topic branch, the commits are combined into one commit and merged into the default branch.

Arash points out other columns like `Acceptance in progress` and `Done` which he will talk about after explaining more about the reasoning behind using *Squash and merge*. He then navigates to the current commit list of the master branch.

"You can see that what I just merged is just one commit, with my custom message as a result of squashing those previous three commits. This is how we want to do things moving forward, and that includes myself, Yousef, and Masoud as well." He scrolls through the rest of the commit list to prove that this practise was followed by the other mentors.

"Of course, it's natural to make mistakes in the beginning and I will remind you again when you do." He assures in a light tone.

After returning to ▥`Ignite`, Arash proceeds to describe the last phases of stories.

"Even though it's in `Merged`, the story is not finished yet. The general rule in software engineering is that developers are careless; They start with some assumptions which are not necessarily always correct, and they may miss things. So, it is always a good idea to have someone else check their work, and that someone could be from your team. I'm not going to do a rigorous test on everyone's code. Most of the time, you will be a developer and your teammate will take on the role of a QA or an acceptance tester and vice versa.[11] It is also best practise to have a third person aside from the developer and the reviewer to do the testing, and once they do, they must assign the whole story that they will be testing to themselves and drag it to `Acceptance in progress`."

Seeing some blank faces on the screen, he clears his throat. "The reason why we need a third-party for testing is because of a common issue that arises is when two people are working on different parts of a task. They may not always check everything together or one person may incorrectly assume that the other is going to be implementing something. These are the situations in which something gets missed and integration does not always work. So, it is good to have a third-party conduct acceptance testing on the complete feature. Bear in mind that acceptance testing is different from actual software engineering testing; It is a more involved process." [12]

He concludes by explaining that once the acceptance tester is satisfied that all the acceptance criteria are met, they can then move the story to `Done`.

---

[11] Software quality assurance (SQA) is a means and practice of monitoring the software engineering processes and methods used in a project to ensure proper quality of the software. On the other hand, Software Quality Control (SQC) is a set of activities for ensuring quality in software products. Software Quality Control is limited to the Review/Testing phases of the Software Development Life Cycle and the goal is to ensure that the products meet specifications/requirements.

[12] Acceptance testing is a formal testing with respect to user needs, requirements, and business processes conducted to determine whether a system satisfies the acceptance criteria and to enable the users and customers to accept the system.

Chapter 5: Sprints

Friday, May 1

"In my early years of iterative development, I thought timeboxes were actually about time. What I came to realize is that timeboxes are actually about forcing tough decisions."

- Jim Highsmith

"I mentioned earlier that sprints should generate deployable code and that we will test a complete story in our demo sessions. I will now briefly discuss what happens in these sessions as well as our process and techniques so that you have an overall understanding of how things will work going forward."

Arash pulls up a new slide in his PowerPoint titled "What Happens in a Sprint":

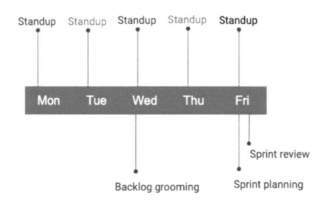

© The Author(s), under exclusive license to Springer Nature Singapore Pte Ltd. 2023
N. Raeesinejad et al., *The Ignite Project*,
https://doi.org/10.1007/978-981-19-4804-6_5

> **A WORD FROM THE MENTORS**
>
> At the end of each sprint, a **sprint review** meeting is held for the Scrum team to show off their completed work. This is usually through a demo of the new features.
>
> Sprint review participants typically include the Scrum team as well as internal and external stakeholders, including management, customers and developers from other projects.
>
> The most important reason for which we have combined our sprint review and planning sessions into one meeting is to save time, as our MEng interns work part-time and should be given enough time to develop features. Another reason is that planning for one-week sprints takes a short amount of time.

"After creating and sizing issues in the sprint planning, we begin the next sprint by working on implementing them. Monday is the first day of our sprint. Throughout the sprint, you will have a couple of mandatory meetings. The undergraduate students will attend all daily standups, while the MEng students will attend only the Tuesday and Thursday ones. On Fridays at noon, all issues that have been tested must be in Merged and completed for the sprint review, which I have mentioned in the last bullet point of the README file."

> 9. Sprint review
>   - Arash will pull the master on his machine
>   - In the Zoom call, each team will get control of the mouse and present their work on Arash's machine.

"I will fetch the latest stable version of the code and with the guidance of each team, I will demo their implemented features." As someone who has mostly worked individually on academic coding assignments and projects to be solely tested on TAs' machines for marking, I'm interested in seeing how a long-term full-stack project like the Ignite system would be demoed on a single machine after being implemented by multiple developers on different systems.

As if reading my mind, Arash suddenly warns, "Having custom configurations that only work for your system and nowhere else does not work. You need to make sure that your code will eventually work on my system as well as with the changes made by the other teams."

He adds that the sprint planning session for the following week will take place immediately after the sprint review. I quickly open my Outlook calendar app and pull up my invitation to the demo meetings. Initially, I was doubtful and not so enthusiastic about a 3-hour meeting, but it is now understandable why it has been planned this way, to accommodate for both planning and demoing.

Arash then declares to the team that they will not do backlog grooming in the first few sprints.

A WORD FROM THE MENTORS

Product backlog refinement or **product backlog grooming** refers to the activities of writing and refining, estimating, and prioritizing PBIs. The goal here is to keep the product backlog clean and orderly and ensure that the appropriate items are prioritized i.e. at the top of the backlog and ready for delivery.

Some of the activities involved with refining the backlog include

- removing user stories that no longer appear relevant
- creating new user stories in response to newly discovered needs
- re-assessing the relative priority of stories
- assigning estimates to stories which have yet to receive one
- correcting estimates in light of newly discovered information
- splitting user stories which are high priority but too coarse grained to fit in an upcoming iteration

Product backlog grooming is a regular and ongoing activity and thus may be scheduled as an official meeting. In a **product backlog grooming meeting**, we ensure the backlog is ready for the next sprint.

Some teams like to hold the product backlog grooming meetings three days before the end of the current sprint; This gives the PO sufficient time to act on any issues that are identified. Other teams may find shorter weekly meetings every week in lieu of one meeting per sprint are more suited to their cadence.

The Ignite team did not hold grooming meetings during the first few weeks as the focus was on training. Since the velocity of the team was low during this time, there was no need for many newly groomed backlog items.

"The idea is that every two weeks, we get together on Wednesday and talk about general things coming down the pipe so that we understand what is coming next. This is so that when we are working on tasks, we have something in the back of our mind regarding how we are going to address issues that may potentially come up later."

The next PowerPoint slide contains a similar timeline but with different labels this time. Arash refers to them as the "technical" aspects of what happens in a sprint.

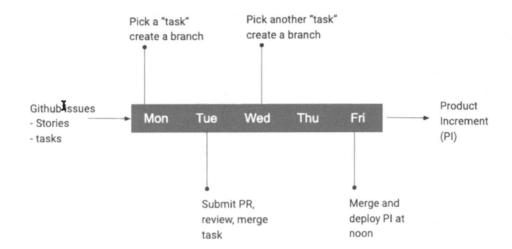

He then defines the product increment for their team as the deployable code in the master and Merged each Friday at noon.[13] He finally concludes all the contribution guidelines, theoretical concepts, and housekeeping items for the team going forward.

"I know this is probably a lot of information to take in at first," he confesses. "Are there any questions?"

"I have a question," Michael declares, off-video.

Michael is a MEng student in Software Engineering at University of Calgary, he graduated with a degree in mechanical engineering with experience in data analysis. He loves basketball and he wishes to work as a data analyst for an NBA team after graduation.

"In the Monday meetings, do we pick stories, or do we pick tasks?" He asks.

Arash answers that they are to pick stories. "Sometime between the planning session on Friday and Monday, the team gets together and decides how to break down their

---

[13] A potentially releasable product is high quality, well tested, and complete but not necessarily cohesive. Being potentially shippable does not mean the results will actually be delivered to customers. Shipping is a business decision; potentially shippable is a state of confidence.

assigned stories into tasks. Starting on Monday, they will pick up those tasks one-by-one and begin working on them throughout the sprint."

Another member by the name of Mihai chimes in, "If we need clarification for a story or a task, who do we ask?"

Mihai is part of the incoming group of students that will be on the Ignite development team. His background is in Civil Engineering and has had some experience in project management and researching/teaching as a grad student.

"If it is about the technology on the frontend side, Masoud is the expert. If it is about whether this feature makes sense or not in achieving the goals we want to achieve, then Mohammad and myself would be the best people to reach. On the technical backend side, either Yousef or myself can answer your questions."

Thanmayee declares she has a question next. "For the first MVP, it doesn't look like any of the tasks are what Kevin and I are supposed to be doing. I was just wondering if we're supposed to be working on a different portion of the project until June or how else is that going to work?"

Thanmayee (Than for short) is a master's student who is currently enrolled in the Software Engineering program at the University of Calgary. With an undergraduate degree in Chemical Engineering complimented by another masters in Materials Science, it is clear that Than loves to learn! She is part of the Ignite project on the Data Analytics and Visualization team.

"As I have said, because of the dependencies between stories, every member will work on whatever is on top of the backlog. It doesn't matter whether it corresponds to what you originally wanted to work on or not." Based on my understanding from previous discussions, each of the three Agile teams would be working on a set of related features according to their preferences which comprise both frontend and backend components.

"However," he notes, "given you have a couple of stories assigned to your team, you are more than welcome to choose the one you want to work on from that list." He suddenly snickers. "Unfortunately, you won't be able to work with data in the beginning, simply because there is no data to analyse yet."

Based on what he said, I'm guessing Thanmayee and Kevin's main preference is to work with data analytics. Although they will not be working on tasks focused on that type of work, Arash adds that they can still work from a data analytics perspective, meaning that as the team is working on stories, they may identify opportunities to gather some data and create stories for adding features for gathering those metrics.

"I would love it if you could gather those kinds of feedback and add them to the backlog to work on later." He requests. Thanmayee nods and thanks him, looking satisfied with his response.

Stephanie is quick to ask her question next, "Could you talk a little bit more on how the stories are assigned, and then how the teams are supposed to work together? I know that our project team is different from our Agile team and I'm just curious about how that connection will play out."

 Stephanie is an MEng student in Software Engineering at the University of Calgary. She has an undergrad in mechanical engineering and spent some time in the digital marketing world as well. She got into the field to build cool things!

I am assuming that the project teams refer to the groups in the team design project courses offered in the MEng program. It seems as if the Agile teams have been arranged such that they encompass multiple project teams.

Arash nods. "So, as an Agile team, you pick the stories you want to do during a sprint. If you can't decide in the first few sprints, I can assign them to you. However, I would rather all of you decide yourselves. Now, picking a task ties with the internal project team as you mentioned. Let's say there is a task involving different features to be implemented. If you are more comfortable working with your project team, you can pair program[14] or split the feature into frontend and backend portions for you and your partner to work on. The point I'm trying to make is that when it comes to tasks, you can choose to collaborate with one person more than another."

---

[14] Pair programming, otherwise known as "pairing", consists of two programmers sharing a single workstation i.e. one screen, keyboard, and mouse. The programmer at the keyboard is usually called the "driver". The other programmer, who is also actively involved in the programming task but focuses more on the overall direction, is called the "navigator". It is expected that the programmers swap roles every few minutes or so.

He checks if he answered her questions, to which Stephanie admits she is still confused about picking the stories. "Are the stories assigned on a first-come, first-serve basis?"

"Well...I-I mean-" Arash stutters in response before coughing to regain his composure. "Ideally, the teams are just going to say which stories they like more and pick them. If more than two teams like the same story in the beginning, I will resolve that conflict with a coin toss, for instance."

Stephanie accepts this answer with a chirpy "Ok!" before playfully warning that she will have more questions down the line. Arash smiles and nods, then waits a couple more seconds for any further questions from the team. Hearing none, he proceeds to announce their plan for the next hour.

"I want each Agile team to get together over Zoom or our Slack channel and break down two stories.[15] The first story is from the training material that I have sent you. You will discuss the story amongst yourselves, come up with its wording, create an issue and size it, break it down into tasks, and then size them as well. Afterwards, you will also do this for the rest of the stories in the training material."

He then lists three user management stories in the <u>release-planning</u> document from which each Agile team may pick one as their second story:

Authentication (social login):

As a user I want to login using my Google account so that I can access the workshop.
- AC1: If a user does not have an account, she can register using her google account
- AC2: If an instructor logs in using her google account, she should see the "instructor" role in her profile
- AC3: Once the user registers, an empty user profile is created
- AC4: As long as the user profile is empty, a warning banner should be shown so that the user can click and update it
- UX: N/A

---

[15] Zoom Video Communications, Inc. (Zoom) is a technology company that provides videotelephony and online chat services through a cloud-based peer-to-peer software platform. Slack is a proprietary business communication platform developed by the company Slack Technologies. It offers many IRC-style features, including persistent chat rooms (channels) organized by topic, private groups, and direct messaging.

> ### Start my JupyterLab (docker):
>
> As a user I want to start a separate instance of Jupyter lab so that I can access the features of the workshop.
> - AC1: When a new user registers, NO user is created for them on the host.
> - AC3: After the user logs in, she can start the JupyterLab from the UI
> - AC4: JupyterLab starts in its own container with the username of the logged in user.
> - UX: N/A
>
> ### Global dashboard:
>
> As a user I want to access a dashboard that will contain information for all authenticated users.
> - AC1: An authenticated user can reach the global dashboard from her dashboard
> - AC2: The global dashboard will contain a greeting message

"You may not have enough time to finish implementation for the next sprint and that is alright, but you still need to break down whichever story you chose into tasks."

Arash then grins and rubs his hands together. "So, who likes what?"

A period of silence ensues.

Slightly deflated, Arash gives the teams five minutes to discuss offline and send him a list of the stories in order of their preferences. He also mentions that after being assigned the second story, they will have one hour to break everything down and size them with their teams.

"Hey Arash, are we doing the stories this week?" Stephanie asks.

"Yes," Arash confirms. "You have to do the training material, but the second story is optional if you have enough time."

Yousef suggests creating breakout rooms in the Zoom call for the teams to discuss which second story they want to work on. Mohammad agrees and starts assigning each Agile team to a different room. In the meantime, I glance away from their videos disappearing one-by-one and review the meeting notes I have been jotting for the past hour and a half. In all honesty, they are subpar at best, despite Arash's extensive explanation of all those concepts which still seem relatively foreign to me. The one thing I know for sure is that to keep up with future meetings, I must increase my knowledge and understanding of this process in theory, before seeing it executed in practise by the teams.

"Yousef, what room should we assign Niyousha to?" I glance back up after hearing Mohammad say my name. "Or should she just float between rooms?"

"I think it's better to have her in one random room." Yousef answers.

After a moment of typing, Mohammad tells me he has assigned me to Team 2's breakout room. Before I leave the main meeting, I confirm with him that I will continue with my general observations of the team's progress.

Chapter 6: Stories to Tasks

Friday, May 1

"One principal problem of educating software engineers is that they will not use a new method until they believe it works and, more importantly, that they will not believe the method will work until they see it for themselves."

- Watts Humphrey

I leave the breakout room and get redirected back to the main meeting. While waiting for the remaining members of the team to return, I scan over the notes that I have written down from observing Team 2 for the past hour. The activity started off a little shaky, with no one stepping up to start off the simulation of creating stories. Most of the members had revealed to have not finished reading the training material. In addition, not everyone had their video turned on and even some had muted their microphones, which I felt made it more difficult to collaborate with each other. Overall, the first several minutes were filled with indecisiveness and hesitance, until one member, Mihai, finally initiated control of the flow of their meeting. From that point, they worked together through his shared screen, providing input regarding what to include in their stories with reference to their training material. Although they were not able to finalize their second story, I believe they have become comfortable enough with the general process of breaking down stories to quickly finish the simulation after this meeting ends.

Arash quickly welcomes back all the team members after returning from their breakout rooms.

"Moving forward, this is how we're going to work on Fridays, except that I will be present and will chair the meeting and ensure that everyone stays on the same page."

He elaborates again that at the end of each sprint, he will present the new features for the team to work on in the grooming and planning meeting immediately after their demo. As a team, they will then discuss the acceptance criteria and size the stories, which they would

N. Raeesinejad et al., *The Ignite Project*,
https://doi.org/10.1007/978-981-19-4804-6_6

start implementing on the following Monday. Arash seems very adamant on not being involved with the creation of team-specific tasks.

"Let's now review the stories you have created."

He shares his screen once again and navigates to ⌨ `ignite` and scrolls through all the issues, their timestamps showing that most of them have been created in the past half hour. The team names assigned to the issues have been changed as well; Per Arash's request on Slack, each team had submitted their own team names during the same time as the simulation. I quickly check the general channel to see the updated team information for myself.

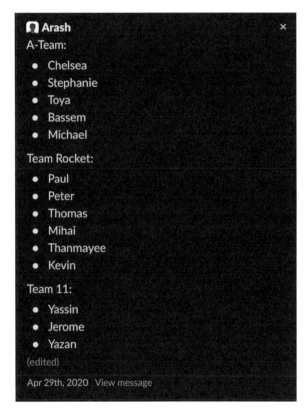

Starting off with the **A-Team**, Arash opens one of their created stories for the training material.

Complete Learning Materials - A-Team #28

Complete the learning material Tasks:

Task 1 - Backend Service - #29
Task 2 - Standalone Frontend - #30
Task 3 - Connect with Jupyter - #32

> Optional Tasks:
>
> Optional Task 1 - Learn Linux Commands - #41
> Optional Task 2 - Learn Agile Point System - #42

Right off the bat, he notes that the first line is not in the correct format and starts adding to the issue description.

> As a grad student, I want to complete the training material so that I know how the frontend, backend, and Jupyter extensions work.

"Stories must also include acceptance criterium," he explains while typing. "When writing them, please try to be as explicit as possible. If we don't specify things like the template, then they could be interpreted differently."

> AC:
>
> - Implement a REST api backend using the given template that has the following endpoints:
>   - Accepts POST requests on /api/echo to save a message
>   - The format of the output is {"message": "message content"}
>   - Accepts GET request on api/echo to get the last saved message
>   - returns {"message": "message content"}
>   - The services run inside docker and responds on port 3000, docker maps the port to 8002 of the host
>
> - Implement a standalone frontend that can communicate with the backend. The frontend should be implemented using React and should look like the image in the training material document.
>   - Once you type a message in the input field and hit submit, it should send a GET request to the django serice.
>   - Every second, it should send a GET request to the django service and show results in the green pane.
>   - Integrate the frontend with Jupyter Lab as the template

"Originally, we wanted to use Typescript, however we have decided that the main portion of the code will be implemented using React and Typescript will only be used when needed."[16][17] Arash refers the team to a folder another repository called ⊟ `sample-dev-templates` to integrate their React projects into Jupyterlab. He also mentions that if the referred image in the AC did not exist in the training material document, it would have to be replicated as a mock-up[18] of the view and added to the issue description.

Arash addresses their list of tasks next, praising the first three main tasks. He suddenly pauses after coming across their list of "Optional Tasks".

"Oh, I see..." he reads them out loud once before commenting, "I think these are nice features, but-" he stops midsentence to ponder for a few more seconds. "Never mind, these are connected to the learning material, so they're good. Good job!"

After confirming their assigned labels, he says they need to size the story. He tries to load scrumpoker on his browser, but it seems frozen. Arash wonders out loud if it has something to do with his internet, to which Toya reveals that his team experienced the same issue with this website. Giving up on trying to refresh his page, Arash tells the agile teams to type their estimates in the chat instead.

A WORD FROM THE MENTORS

In planning poker, when the story has been fully discussed, estimators privately select one card to represent their estimate. All cards are then revealed at the same time.

If the cards are not shown at the same time, the formers' estimate may affect the next. As such, there is a risk of biased estimations within the Ignite team!

Within a few seconds, almost all the members post a unanimous "5" in the chat.

"Looks like no debate is needed." Arash notes with a satisfied smile. After assigning the appropriate labels, he declares the story ready. This is the exact workflow for all planning sessions going forward, however the teams are now expected to ask further questions

---

[16] React (also known as React.js or ReactJS) is an open-source JavaScript library for building user interfaces. It is maintained by Facebook and a community of individual developers and companies. React can be used as a base in the development of single-page or mobile applications. However, React is only concerned with rendering data to the DOM, and so creating React applications usually requires the use of additional libraries for state management and routing.

[17] TypeScript is an open-source programming language developed and maintained by Microsoft which builds on JavaScript, one of the world's most used tools, by adding static type definitions. Types provide a way to describe the shape of an object, providing better documentation, and allowing TypeScript to validate that your code is working correctly.

[18] In manufacturing and design, a mock-up is a scale or full-size model of a design or device, used for teaching, demonstration, design evaluation, promotion, and other purposes. It is a prototype if it provides at least part of the functionality of a system and enables testing of a design. Mock-ups are used by designers mainly to acquire feedback from users. Mock-ups address the following idea captured in a popular engineering one-liner: You can fix it now on the drafting board with an eraser or you can fix it later on the construction site with a sledgehammer!

and contribute to the AC. He expresses his hope for the team to not be passive in being told what to do but be more active in communication.

Arash proceeds to briefly scan over the other teams' stories, careful to not repeat the entire process he just showed. I jot down some small additional remarks that he makes along the way:

- *You can also explain what the feature looks like from the perspective the product owner, if not the end user. Both are valid views.*
- *AC of user story can change depending on who we define as the user.*
- *The point of AC is to be tested and verified (cannot be something abstract that cannot be measured)*
- *Do not assign story point labels to tasks*

"It is probably hard at first for you guys to break down stories into tasks because you are not familiar with the system and that is completely alright," he sympathizes. "Once you are done with your first story and still have time during this sprint, pin me on Slack and I will describe the system, such as how it works and how you can replicate the development environment on your machine, and how to start Docker."

Arash announces that he will review Team Rocket's optional story next, to which Mihai quickly brings to his attention that they did not have enough time to break down their story into tasks. Unruffled, Arash repeats the same directions, telling them to finish that activity and reach out to him afterwards to practice breaking down more stories.

"Our optional story is complete." Yassin says.

Yassin is a PURE research student on the Ignite Team. He is entering his third year of Software Engineering at the University of Calgary. His focus is on deployment and back-end development. Outside of programming, Yassin likes to play soccer and video games.

Arash proceeds opens Team 11's optional story.

Global Dashboard Implementation - Team 11 #40
As a user I want to access a dashboard that will contain information for all authenticated users.
AC:
- Authenticated Users will have access to the Global Dashboard
- User should be presented with a greeting message

Tasks: - #43 - #44

After verifying that their labels are correct, he looks at the first of their two tasks.

> **Make the user interface for global dashboard - Team 11 #43**
>
> Time estimate: 1 day
>
> Create the dashboard using HTML.
> Style the dashboard with CSS.

"Don't just use HTML and CSS, try to use React-" Arash suddenly refrains himself from explaining further, saying he will discuss React in a later session. He then reviews their next task.

> **Connect Global Dashboard with User Dashboard - Team 11**
>
> Time Estimate: 2 hrs
> Using Typescript, have buttons to navigate between Global Dashboard and User Dashboard

After quickly going over the content of the task, Arash notices a question posted in the chat.

> Stephanie: **Do we have training materials for React?**

"Masoud will probably post them since he is the expert on React."

> Stephanie: **Ok thanks!**

"Regarding the time estimates for tasks, would it better to use hours or days?" Peter asks.

"It depends on you," Arash replies. "Personally, I try to define tasks that are at most a day. A task that takes more than a day means it is not well-defined."

"You should estimate tasks based on hours." Yousef adds.

A WORD FROM THE MENTORS

It's useful to estimate both the product backlog and the sprint backlog because the estimates are used for different reasons.

Product Backlog Estimation

Product backlog estimates are typically expressed as story points or ideal days. This kind of estimation typically occurs in product backlog grooming meetings.

Sprint Backlog Estimation

Tasks in the sprint backlog are typically sized in ideal hours (also referred to as effort-hours, man-hours, or person-hours). For instance, if the team estimates that a task will take 5 ideal hours to complete, that doesn't mean it will take 5 elapsed hours; It might take one person several days or it might take less than half a day if several people are working together.

There are 2 reasons to estimate the sprint backlog:
- ✓ It helps the team determine how much work to bring into the sprint.
- ✓ Identifying tasks and estimating them during sprint planning helps team members better coordinate their work.

Arash summarizes the planning session for the teams one more time, noting that everyone must understand and agree with the acceptance criteria and size points of a story before breaking it down into tasks. He then tells the teams to identify any of their recently created stories and tasks that they think they can do for this sprint and drag and drop them into the project board. He waits in silence for the teams to do this.

"I just have a brief question," Bassem announces. "Are we meant to put just the stories in the board or everything, including stories and tasks?"

 Bassem is a software engineering Master's student. He has a mechanical engineering background and several years experience working in downstream natural gas project engineering. He is currently pursuing a lifelong dream, a career in video game development.

Arash confirms it is the latter, noting that in the development cycle, the teams will be working on tasks which will be advanced through each column in the project board. It follows from this that stories are never assigned to one member and they must move in the To do column. Once all the tasks for a story are finished, then that story will be closed either automatically or manually.

Seeing that there are no more questions about the workflow so far, Arash reverts to an earlier inquiry from one of the members about what to do if more than multiple people are working on the same task.

"Do whatever is best for your team, whether that entails assigning more than one person to a task or creating duplicate tasks. The specific name assigned to a task is not that important."

He switches back to his latest PowerPoint slide and verifies that they have covered everything in their plan for today.

"Starting on Monday, you will all be working on stories and tasks for the training materials. If you have any questions along the way, pin any of the mentors as well as myself. We can pair program and share our screens to work on any issues you may have."

Arash also points out some resources to help the teams get started. The first resource is another repository called ⊟Sample Dev Templates. He explains that the team is supposed to modify two files called django_rest_api and Jupyterlab_react for their training material.

"You will add a new service to the services directory in our ignite repository. You can name it something like *training material-A-Team,* for example. This service will also contain a backend and frontend component. The backend folder should contain a copy of the django rest api and for the frontend, it will be the Jupyterlab_react file."

He voices his premonition that the team will have several questions once they start working on this training material, including how Docker works, how to compile all their code, how to connect Jupyterlab, etc.

"Whenever you hit any of these walls, let me know and I will be happy sit down with you and show you how it is done. If you want to work on it on your own, I encourage you to look at this."

He opens a folder in ⊟ignite called how-to. Inside, he points out a file called setup. He highlights a section in the file that covers how to run a chat in Jupyter.

"Just to give you a heads up, developing on Jupyterlab is slow. Instead of developing on Docker inside Jupyterlab, create a Conda environment and do it on your own machine.[19]

---

[19] Conda is an open-source, cross-platform, language-agnostic package manager and environment management system. The Conda package and environment manager is included in all versions of Anaconda and Miniconda. The main purpose of Python virtual environments is to create an isolated environment for Python projects. This means that each project can have its own

However, when it comes to developing the Django standalone, for example, you can use Docker. Are there any questions?"[20]

"Umm I have a question," Chelsea asks, off-video. "For the training material, does every person need to add a service folder or is it per team? Because it doesn't make any sense for me to pull something that is half-way done to learn."

Chelsea is a Master of Engineering (M.Eng.) student in Software Engineering. She has an undergrad in Civil Engineering and worked as a Transportation Engineer for 15 years prior to starting the M.Eng. program. Chelsea's first program was written on an Apple IIe in BASIC.

"Sure," Arash shrugs. "You can have your own username as the prefix instead. Good question."

He waits for a few more seconds. Seeing as no one has any more questions, Arash declares that his presentation is over.

---

dependencies, regardless of what dependencies every other project has. Users can create virtual environments using one of several tools such as Conda.

[20] Django is a high-level Python web framework that enables rapid development of secure and maintainable websites. Django's primary goal is to ease the creation of complex, database-driven websites.

Chapter 7: Learning Curve

Friday, May 1

"Anyone who stops learning is old, whether at twenty or eighty. Anyone who keeps learning stays young."

- Henry Ford

Following the end of Arash's presentation, Mohammad first thanks everybody for sitting through the kickoff meeting. "There's lots of new information to digest," he sympathizes. "In the first couple of weeks, there will be quite a bit of a learning curve, but this is all going to make sense and you will have a smoother ride. During the first two weeks, we are expecting a lot of questions and so we are planning on pair programming with your teams to make sure that you are on the right track."

He confirms that he will send the meeting invites for all Friday meetings and daily standups and encourages the team to accept these invites and keep them in their calendars going forward.

He specifically addresses the MEng students next with an amused smile. "Hopefully you are beginning to feel that this is not just a course and that you are actually part of a company. Back in September, I had conversations with some of you about how overwhelming things were and look at you now, basically software engineers! At this stage, as we are trying to figure out which processes to put in place, let's use the next couple of minutes to reflect a little bit. What did you guys think about the meeting?"

"This is like being back to week one of bootcamp." Chelsea admits.

"Welcome to Software Engineering!" Mohammad laughs. "People who are going to be successful software engineers are those who are agile in learning new things. And most of the jobs you have will not expect you to contribute to production for at least a couple of months. So guess what? Another bootcamp will roll out! I think it's really cool that-"

N. Raeesinejad et al., *The Ignite Project*,
https://doi.org/10.1007/978-981-19-4804-6_7

His voice cuts off. Startled, I glance up from my notes and am met with Mohammad's frozen grin.

After a period of awkward silence, Arash asks if Mohammad is still on the call but his only reply is some static noises. After another couple of seconds, Mohammad's audio and video are finally restored.

"Do you guys hear me?" he asks in a flustered tone.

"We do now." Chelsea replies.

"Ok," he chuckles, "What was the last thing you heard?"

"*I think it's really cool that...*" Chelsea mimics and a couple of people snicker in the background.

"Right! I think it's really cool that we're experiencing this together." Mohammad pauses to read a new comment in the chat.

Paul: It was really helpful to walk through an example of the GitHub workflow.

Paul is one of the Software Master's students on the Ignite Team. He is a BSc in Mechanical Engineering trying to switch to a career in software or game development. For Ignite, his primary focus was front end development for the live lesson component of the application.

"I agree with Paul," Jerome says, "Working through the actual sprint process helped a lot with understanding it."

Jerome is one of the undergraduate students working on the project. He just finished his second year of software engineering. He was introduced to the project by Dr. Moshirpour who taught some of his classes and judged a coding competition he participated in.

Mohammad nods, "It's really important to make sure that you're comfortable with the terminologies."

"Would it be unreasonable to ask what your expectation is of us to make it through this bootcamp and actually be productive?" Chelsea inquires.

"In terms of time?"

"Yes, because we just finished our exams and with the new semester starting next week, this seems much more compressed than what we are used to," she reveals. "While I understand that this is the norm in industry, we wouldn't be expected to start our bootcamp until we actually started our work. There's a slightly different dynamic here that I would like to get a better sense of how quickly we would be expected to go through that learning curve and produce usable code."

"Of course. The idea was to have the training material basically done by next Friday." Mohammad scratches his head. "It's tough to predict exactly how things go, especially since each person progresses differently, so we are going to just take it one sprint at a time. We'll see how people are doing and how much support they need. By next week, we will start working on some features little by little. The key thing in this whole Agile process is to think one sprint at a time and learn from our progress from previous sprints to scale ourselves for future sprints."

"There isn't really a fixed time limit for solely learning," Arash adds. "Of course, there is a learning curve in the beginning just to see how everything is set up and we don't expect anyone to be an expert in everything in a couple of days. You will learn more by completing more tasks along the way, and the number of tasks is adjusted based on your learning rate for each sprint."

A WORD FROM THE MENTORS

A **learning curve** is a correlation between a learner's performance on a task and the number of attempts or time required to complete the task; this can be represented as a direct proportion on a graph. The learning curve theory proposes that a learner's efficiency in a task improves over time the more the learner performs the task.

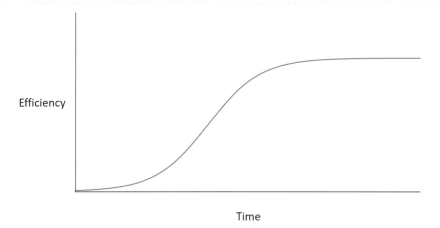

The S-Curve or Sigmoid function shown above is the idealized general form of all learning curves, with slowly accumulating small steps at first followed by larger steps and then successively smaller ones later, as learning efforts reaches a plateau.

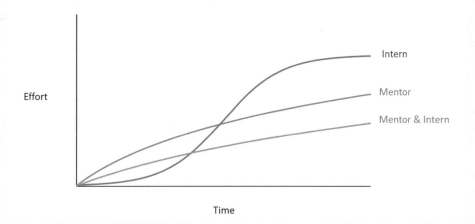

From the Ignite team's perspective, hiring student interns is a means to decrease the amount of effort from our experienced developers. At a certain point in time, the interns overcome their initial steep learning curve and eventually become more productive, generating more value compared to the effort of experienced developers if they were to build the Ignite system on their own without the contribution of interns. The experienced developers instead act as mentors to interns which would mean they would lose some productivity because of the time they would have otherwise invested as developers. As time progresses, everyone's productivity increases!

Mohammad nods. "One good example here is becoming good at estimating. Let's say you want to pick a bunch of tasks to do, and you don't want to bite off more than you can chew but you also don't want to not have enough work. This is another learning process for everyone to reflect on."

"In the beginning, will you help us with estimations?" Stephanie requests. "I know that when I started as a junior Mechanical Engineer, we weren't told to figure this out ourselves, but instead that we have a specific number of hours to do something. We don't have any real idea towards how long it takes things to happen in software. We may have done assignments but they're not real."

"The problem with estimations in software is that no one knows how to do it. Even I don't!" Arash laughs.

Mohammad agrees that the Agile methodology is certainly different from traditional Engineering methodologies. "It's quite innovative. Obviously, the more estimations you do, the better you get at them. But even experienced Software Engineers experience significant fluctuations in estimates when starting new projects."

"Estimations are always wrong," Yousef says with a knowing smile. "But the real question is: Why do we estimate in the first place?"

Arash and Mohammad nod, sharing the same knowing smile, but remain tight-lipped. I'm guessing Yousef's question is purely rhetorical though I have not yet fully grasped the real reason behind estimations.

A WORD FROM THE MENTORS

One of the main reasons for estimations in project management may be broken down into its long-term vs short-term benefits:

Short-term benefits
Finding misunderstandings amongst team members regarding the scope of the problem and solution. More important than merely projecting the number of days or hours required to complete a task, the real reason behind estimations is to assure that everyone the team is on the same page at every sprint.

Long-term benefits
Monitoring the trends of story consumption and detecting and reacting to potential problems. This is otherwise known as **velocity**, which is a simple measure of the rate at which work is completed per unit of time by the team. Using Scrum, velocity is typically measured as the sum of the size estimates of the product backlog items that are completed (unit of finished work)

within a sprint (unit of time). So, estimating stories allows to measure and monitor velocity as well as react to low velocity.

In the context of the Ignite project, since we were working with inexperienced interns and only had them for a short period of time, we were not looking at the long-term objectives of estimations but rather were targeting its short-term benefits.

Mohammad reads another comment from Paul in the chat.

Paul: One thing I feel like could have been better communicated is unit testing projects.

"Technically you will have a course on testing in the Summer, which is a little ill-placed," Mohammad admits. "Unit tests aren't silver bullets and as such, they won't have a large role in this project." He prompts Arash to further explain their reason behind not incorporating a lot of unit testing.[21]

Arash is quick to answer. "It depends on our experience and time as well as the value we get out of it. Personally, I usually write unit tests before actual code, because it gives me a better idea of how to define my functions and interfaces. The problem is that this slows down development, especially in the initial phases. If we have limited human resources and time, and we must decide between testing at the function level versus the high-level features. The latter is more important. However, if we didn't have these constraints, then we should test at both the function level and integration level. Given our team's time commitments and limits, we have decided to only conduct higher-level tests to verify that the entire system works."

"Exactly," Mohammad says. "And this is a great example of architecting your project based on your needs. You guys should absolutely know the theory behind testing as well as its practical use. However, even testing has a place or no place in particular projects."

He moves on to read the next question posted in the chat.

---

[21] Unit testing is a level of software testing where individual units/components of a software are tested. The purpose is to validate that each unit of the software performs as designed. A unit is the smallest testable part of any software. It usually has one or a few inputs and usually a single output. In procedural programming, a unit may be an individual program, function, procedure, etc. In object-oriented programming, the smallest unit is a method, which may belong to a base/ super class, abstract class, or derived/ child class.

Peter: One more question for Arash - I know you mentioned we will be using more functional testing rather than unit testing. Could you expand a bit on the process for which we should follow the functional testing?

"Suppose you have finished a task for the training material feature we discussed earlier," Arash starts explaining. "The backend[22] is supposed to respond to certain **GET** and **POST** queries. A test that is sending a query to your endpoint[23] and receiving the correct value from it is testing that endpoint at the functional level. Now, what happens under the hood inside that code - whether one function or a hundred different functions are being called - doesn't matter at that level. As long as I get the correct value out of it, it's fine. That is the level of testing I am talking about. Moreover, we will not have automated tests for the UI, but just at the API level."

To that, Peter unmutes his mic to ask a couple of follow up questions. "Who will be performing those tests? Should that person be a third party? Does there exist some sort of testing process document that we should follow?"

"Yes!" Arash nods vigorously. "However, to avoid overwhelming you from the very beginning, we have decided not to talk about testing and let you just do the work."

Arash

The work we did in the ignite project was quite successful and all the stakeholders benefitted from it. But the were some decisions that we made early on which turned out to be not as productive as we had hoped; Not introducing the students to the testing practices from the beginning was one of them. Initially, I believed that given the

---

[22] The "backend" refers to any part of a website or software program that users do not see. It contrasts with the frontend, which refers to a program's or website's user interface. In programming, the backend is the "business logic and data access" layer while the frontend is the "presentation layer."

[23] Simply put, an API endpoint is the point of entry in a communication channel when two systems are interacting. It refers to touchpoints of the communication between an API and a server or service. The endpoint can be viewed as the means from which the API can access the resources they need from a server to perform their task. An API endpoint is basically a fancy word for a URL of a server or service.

student's backgrounds, it would be too much to ask them to learn two new languages (Python, React), new concepts (REST Api), software engineering practices (such as working in an Agile team), and on top of that, writing tests. Therefore, my approach was to limit their learning to small scopes as much as possible and to make it easier on them, I removed the need for automated testing in the first month.

The students stood up to the challenge and learned all those concepts in the first two months. But introducing automated testing in the third month turned out to be quite problematic, since it went against the norm to which they had already adapted. Learning from this, in future projects I will put more trust on the learning capacity of the new students and establish the proper automated testing practices from the beginning of the project.

"In probably two to three sprints, your PRs must include both your code as well as the test for it.[24] We will have automated code running on GitHub so that whenever you submit PRs, it will run all the tests to ensure that they are all being passed. This is how we are going to make sure that the new feature will not break an old feature, since the automated test of another feature could break if there was a conflict. Other than that, we have manual testing. This is shown by our project board which has columns for `Merged`, `Testing in-progress`, `Done`, and `Accepted`. For `Testing in-progress`, someone is doing manual testing for another team member."

After Peter thanks him for his response, Mohammad concludes the meeting with some parting words. "Anytime you start a new project, there is a learning curve involved - growing pains and all that. People in MEng are definitely unafraid of challenges, that is why they came to MEng after finishing a bachelor's degree outside of Software Engineering. And the students in Software Engineering obviously sought out a new challenge when applying for the PURE award. The product, although very important, is secondary to you solidifying your Software Engineering skills. I would like to encourage all of you to take ownership of your own learning and development and understand that our biggest goal here is to make good legitimate Software Engineers."

---

[24] Contributions to a source code repository that uses a distributed version control system are commonly made by means of a pull request (PR), also known as a merge request. The contributor requests that the project maintainer "pull" the source code change. If the maintainer accepts the PR, they must then merge that contribution to become part of the source base. A comment thread is usually associated with each PR, allowing for focused discussion of code changes.

Mohammad

I have been recently considering incorporating the agile process in my undergraduate SE courses, where my TAs could play the role of a coach or mentor. It would be beneficial for students to be shown the proper process that is being used in industry. However, Scrum doesn't really work for assignments because you cannot enforce anything; at the end of the day, students just receive a grade. Regardless, it remains an issue that students are only being told to meet requirements and not necessarily how.

Teaching the process in a SE course could also take away from the programming concepts. It may be difficult to enforce Scrum, but it would be incorrect for me to say my courses incorporate project-based learning if the proper process and tools are not at the very least introduced. After discussing with experts from both academia and industry, I learned there are certain elements of Scrum such as timeboxing, rhythm, and demos which may be inherently integrated in a SE course outline without direct implication of their enforcement. This way, we can encourage more student planning through activities like breaking down stories and assigning and prioritizing tasks.

What makes developing the Ignite system different from students' courses is that there are heavier implications due to the resources invested in such a large system. It is actually often the case for new company hires that not much is expected from them in the first 3-6 months of their work term to allow for them to learn. Due to our time crunch, one of my main goals is to make the students less stressed and eliminate any hint of imposter syndrome that may arise as a result of their anticipated steep learning curve.

Section III. Training Week

Chapter 8: Standups

Monday, May 4

"Daily scrums can be pivotal to the success or failure of using Scrum"

- Mike Cohn

8:25 a.m.

Closing the door, I quickly grab my headphones and walk over to my desk. Setting down my steeping hot cup of green tea, I plop down on my chair, propping open my laptop and notebook. After typing in the meeting URL, I'm prompted to wait until the host - Mohammad - starts the recurring Zoom meeting. In the meantime, I connect my headphones, put on my glasses, and check if my mechanical pencil is filled to the brim with lead, all while mentally counting down the minutes until 8:30.

It is crucial to have everything ready and perfect for the first ever standup meeting. I may be going overboard, but I want to make sure to not miss anything important during the meeting.

Suddenly, the Zoom window closes and reopens, and I'm immediately greeted with the faces of Mohammad, Yousef, and Arash. Everyone else I.e., the PURE students and myself, are muted and off video.

"Ok Team 11, we're going go over each member's update on what they've done and what their plan is for today." Arash states briefly.

Each member in Team 11 then proceeds, in an orderly fashion orchestrated by Arash, to unmute themselves and give their updates.

"I'm still going through the learning material and will continue to do so today..." Yazan trails off.

"Have you assigned a task to yourself?" Arash asks.

"Not yet, I can do that today."

I quickly jot that down while Arash nods for the next member to start.

"I'm planning on continuing to go through the training documents. I haven't assigned a task to myself yet either." Jerome says.

Slowing down my hand, I glance up to catch another nod from Arash.

The last member, Yassin, echoes the same update.

"Ok, thanks everyone, talk to you soon."

And just as quickly, I'm faced with my desktop wallpaper again. My eyebrows rise when I see the clock.

8:40 a.m.

I look down again at my originally blank page containing the two sentences I had messily jot down in my previous urgency:

Update on which task each team member is working on. Members are still getting accustomed to assigning tasks to themselves on GitHub project.

Huh. Standup meetings are a lot shorter than I thought.

Tuesday, May 5

"Ok let's start." Arash announces and I crack my knuckles while waiting for him to load the Ignite project board on his shared screen.

My initial shock towards the actual length of a standup has gradually died down since yesterday morning, and while I'm not expecting this standup for the Master students to be any different, I still prepare myself with the aim of transcribing more notes than my lame 2-sentence recap of the last meeting.

Arash begins with his own update. "I worked on this Python Django guideline and submitted the PR. I'm not going to finish it, but I will talk more about this after the standup. That's it for me."

He switches to his Slack tab showing a pinned message containing a list of all the teams.

"I guess I need to change this," he highlights the first team's name on the list, written as "Agile Team 1". It seems he has forgotten to change the team names to their official ones established during the last meeting. Verifying that the first team is now referred to as the A-Team, Arash proceeds to call out their first member, Chelsea.

"I-I'm not terribly sure I understand what you need from me right now," she admits.

I would've probably said the same thing if I were in her position. From my very basic understanding, standups are for every team member to give a brief daily update. I imagine an office meeting room setting where each person in the team stands up to say a couple sentences and sits back down.

"In a standup, you need to describe what you plan to do today and what you have done previously in one to three sentences. No details. And if you need help with anything, this is the time to ask for it." Arash explains.

A WORD FROM THE MENTORS

On each day of a sprint, the team holds a meeting called the "**daily scrum**" or "**daily standup**". A daily scrum meeting is ideally held in the morning, because it helps set the context for the day's work.
The daily scrum is timeboxed to no more than 15 minutes.
During the daily scrum, each team member answers the following three questions:
- What did you do yesterday?
- What will you do today?
- Are there any impediments in your way?

The daily scrum meeting is not used as a problem-solving or issue resolution meeting. Issues that are raised are taken offline and usually dealt with by the relevant subgroup immediately after the meeting.

"Umm," Chelsea starts hesitantly, "So yesterday, I started working on Django and the training material and today I plan on continuing to work on the training material."

"Awesome, and have you assigned your name to a task and moved the task here?" He hovers his mouse over the `in-progress` column in the Ignite board.

"... Not even close."

"Please do that," Arash reminds. "You don't need to finish something, you just need to have your name on a task, put it `in-progress`, and then continue working on it. That is the first thing that happens."

"Sure."

"Thank you. Alright, Stephanie."

Stephanie reports in a light tone, "Yesterday I started looking at React - trying to build the little chat for the UI - and then today I'm going to continue doing that!"

"Awesome, and same question: Do you have a task assigned to yourself?"

"No, but I will do that today."

"Please do, thank you. Toya."

"Yesterday I built the Django Rest framework," Toya reports, "but realized I didn't actually build the learning requirements one - the one you had in the sample development templates. So today I just want to actually build the learning environment's Rest framework and..." After a pause, he finishes with a soft "yeah".

"Awesome. Bassem?"

"Yesterday, I kind of went through all the materials. I haven't started yet but that's what I plan to do. I'm going to start on the backend service but I'm pretty familiar with Django, so I expect to finish that today. And then I'm going to start on the standalone frontend."

"Perfect. Michael?"

"I downloaded the Docker toolbox, and I built the image yesterday. I'll plan now on assigning myself to a task to complete the Django and Rest training material."[25]

With each member's update, I'm finding myself increasingly overwhelmed and impressed by the many different topics that are being included in the training material. It's fascinating, given my lack of knowledge in all these mentioned tools and languages, and I almost envy the other students for having the opportunity to learn about them.

Arash moves on to **Team Rocket**, otherwise labeled as "Agile Team 2" on his screen.

Paul is first to report. "So yesterday, I mostly finished up my Django background, and now today I'm going to try to implement the Rest of the API for Django."

---

[25] Docker Toolbox provides a way to use Docker on Windows systems that do not meet minimal system requirements for the Docker Desktop for Windows app.

"Alright, so please assign a task to yourself and move it `in-progress`," Arash reminds again.

"Will do."

"Thank you. Peter."

"Yesterday I started the course on React and today I plan to continue looking at that. And I haven't assigned a task to myself yet." He then quickly adds, "I've got it written down that I need to do that." It's almost funny how everyone is now walking on thin ice around Arash regarding task assignments. I'm not sure whether it was an explicit requirement or not for them to have this done before the standup, but they seem to have picked up on Arash's frustration that it wasn't done.

"Thomas."

"Yesterday, I completed the Rest Django API, and then today I looked at Docker and JavaScript."[26]

Thomas is one of the members on Team Rocket which focuses on front end development. He has a background in Chemical Engineering. In his spare time, he enjoys playing the piano and volleyball.

He takes a pause before hesitantly adding, "and I still have to assign myself to a task." That last phrase is slowly becoming a mantra for everyone.

"Alright," Arash sighs, "Mihai?"

"I got the backend done, I did the frontend in JavaScript and in React, but," he pauses before continuing, "I don't really have any good resources for doing it in Typescript so I might need a little bit of help with that."

"I will talk about that at the end," Arash assures. "We don't need Typescripts."

"Oh okay, because I know the JupyterLab extension-"

---

[26] Representational state transfer (REST) is a software architectural style that defines a set of constraints to be used for creating Web services. Web services that conform to the REST architectural style, called RESTful Web services, provide interoperability between computer systems on the internet. Web service APIs that adhere to the REST architectural constraints are called RESTful APIs or REST API. The Django REST framework is a toolkit for building RESTful Web APIs.

Arash interrupts, "Yeah, that was my bad. It was miscommunicated. I will talk about that."

"Alright," Mihai accepts. "And about the tasks in GitHub ... my team just kind of put them up. Can we delete it and have separate ones for each individual person?"

"Sure. You can change one of the tasks for one person and add more tasks. I think you can delete issues. Ok, Kevin?"

"I have the backend ready, I believe, and I'm just working on the frontend right now." Kevin hesitates before continuing. "Umm... I've been running to a few issues mainly I don't know where to start. I think it would really help if we can talk about it and I can ask questions later."

"We can definitely talk more about this after the standup. Yassin?" After a moment of silence, he clarifies, "Team 11, Yassin?" More silence followed. "Is Yassin on the call?"

"I don't see him," Mohammad says, "I'm not sure what happened. We can move on to Jerome."

"Ok, sure. Jer-"

"I think I got skipped!" Thanmayee quickly interrupts. She is also part of Team Rocket.

"Oh yeah sorry about that!" Arash grins, embarrassed.

"No worries," she assures, smiling back. "I've been working on learning the frontend materials. Also, I've been working on the backend and frontend concurrently, but I'm still having a lot of issues with getting Docker up and running. I'm having the same issues as I was having yesterday. I'm not really sure how to fix that."

"We will talk about that as well," Arash promises, jotting something down on the side. "Jerome?"

"Yesterday, I started working on the Django Rest material, and then today I'm hoping to finish that up," Jerome reports.

"Alright." Arash then moves on to the mentors' updates. "Masoud?"

Masoud is an M.Sc. student in Software Engineering at the University of Calgary. He is currently working on the efficiency and performance of machine learning algorithms.

"Yesterday I revised the React learning material," Masoud says. "The guidelines are already in. Today I'm working on creating some learning materials." I'm slightly disappointed that he doesn't go into detail regarding which topics will these new learning materials be covering.

Yousef follows with his report. "I didn't do too much, but today I'm going to write some sample code for Django patterns."

"Thank you, and Mohammad, do you have anything?"

"Yes, so I think Yazan is going to join Team 11, at least for the time being. We sort of have to talk about setting him up and everything."

"Sure, we can do that."

Yazan is one of the developers on the Ignite Team. He is an undergraduate student, pursing a degree in both Software and Electrical Engineering at the University of Calgary. He has been involved in working on the frontend, backend and deployment of the software. His passion includes Algorithm Design, AI and machine learning as well as Radio Frequency Engineering.

"He's on the call right now so he can give an update."

Yazan's video appears on the screen, to my surprise. Almost all the previous members have been off video so far.

"Yesterday I worked on the Django material for getting the backend and I'm at the very end of finishing that. I will go and do the frontend stuff after that."

After Yazan's report, Arash declares that the team has reached the end of the standup. "Since we have a couple of topics to talk about and we have some time, I'm going to go over them. If we go overtime, we can take the discussion offline."

"Hey Arash, do you mind if I interrupt you for a second?" Bassem asks.

"Yeah, sure."

"I'm with the A-Team and we were just wondering, if it's possible, if we could just have one person assigned to each task and then once the entire group finishes that one task, we knock it off the list?" After a pause, he adds, "And we're only going to do that for the training materials. After that, we're just going to split up the tasks."

"It's alright," Arash responds, "but when you are in a standup, let me know which story it is so that I can see its status."

"Perfect, thank you."

Arash adds, "The whole purpose of having the board is that when we are looking at it, we want to make sure that everyone has something to do and check the status of each task. It's mainly for communication purposes."

He then starts to address each of the team's technical questions one by one.

"We have a question about Typescript and one for Dropbox on Windows. I think Kevin also had a question, and I want to talk about the pull request I submitted." He proceeds to explain that he will start with the pull request, which will supposedly cover the Typescript part, and then he will address the Docker issue.

"Since this is all pretty new to you and it's hard to grasp everything in a day or two, I have done this task."

He pulls up a new issue called ⓘ`Django python guidelines.` "Whenever you need to see what is going on and what we need to do, you can look at this as a guide. I have split it into different commits so that you can understand which code change is required for different things."

Opening a terminal, he proceeds to demonstrate the contents of the backend and frontend of the training material, explaining the portions included in the new guidelines for the team to review for reference.

Returning to GitHub, Arash reminds everyone, "Please commit often and push your changes at the end of the day, so that I know you are in the right branch and what you're doing. This way, I can catch mistakes sooner. So, don't wait until the end of the sprint to push your all your changes."

The next part of his PR is about the Django Rest API. Arash explains it has an authentication section that is not useful, hence why he has removed it, and urges others to do the same in their examples.

Arash's final PR is an Echo service. He tells the team that they don't need to write any migration parts for the API.[27] "As you learn Django, you'll find that the migrations are generated automatically. So, in PR reviews, we can simply ignore them." He lists some other classes for the Echo, such as the model, serializer, routes, and view.

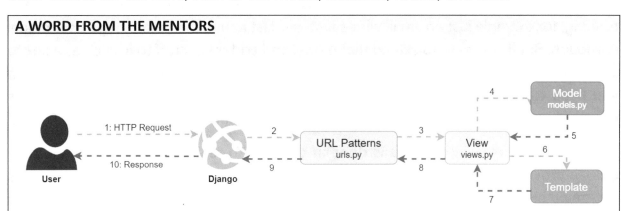

A WORD FROM THE MENTORS

The Django architecture is based on MVT (Model-View-Template) i.e. a software design pattern for developing web applications.

The Model Layer
The model is the single, definitive source of information about your data. It contains the essential fields and behaviors of the data you're storing. Generally, each model maps to a single database table and written in a "models.py" file.

The View Layer
Django has the concept of "views" to encapsulate the logic responsible for processing a user's request and for returning the response. A view function, or view for short, is a Python function that takes a Web request and returns a Web response. The view is usually written in "views.py" file.

The Template Layer
The template is a text file which can generate any text-based format (HTML, XML, CSV, etc.). It contains the static parts of the desired HTML output as well as some special syntax describing how dynamic content will be inserted. To access the view via a URL, we need URL mapping which is done in Django by editing the project's "url.py" file.

"Once you know Django, it will be pretty easy to add these services, but I understand that getting to the point of knowing Django takes some time."

---

[27] Migrations are Django's way of propagating changes you make to your models (adding a field, deleting a model, etc.) into your database schema. They're designed to be mostly automatic.

"One question," Toya cuts in, "I noticed that in your Rest API example and also this one, you have a user model. Why do you make a custom user model when Django already has authentication users?"

"Great question. That is for future extensions. If you want a simple application up and running, for example just an email with a first and last name, then the Django user model is enough. But if you want to extend that model and add other stuff to it, such as a link to a profile or receive more information after logging in, so having a more robust user, then you need to have your own custom user. This is not a completely separate user. It is effectively overriding Django's user and gives us the option of extending it in the future."

He proceeds to explain how this is done in detail, with reference to some of the User classes provided by Django. Toya follows up with another question regarding Arash's reasoning behind using a particular User class, to which he admits that at the time, he had done some research and found that it was the proper class to extend, but does not remember the exact answer. Toya accepts his answer.

Arash moves on to address the next question regarding TypeScript. He pulls up a folder named jupyterlab React, explaining that it contains the format of the Django extension. He points out that certain files have been written in TypeScript and implies for the team to have sufficient knowledge in the language to be able to read and understand such files. To implement a new UI, the team may use most of those files, with some added lines of code. Otherwise, everything else will be done by a widget file containing TypeScript code.[28]

"This is why I said that knowing TypeScript isn't that important. You have a starting point, which is already built for you, and you will change some string and class names. But after that, all of your code will be in React."

He addresses Kevin's question next pertaining to some technical difficulties that arose amidst installations.

"If you can install the Windows Dropbox, then things should mostly work without any problems. The only thing that you would need to change is one line." He shows how to add this necessary change to work on Windows by changing a string inside a Docker file belonging to one of the backend folders.

---

[28] In web design, widgets are small web components that expand the functionality of a webpage or website. In this context, they are JupyterLab widgets, which are distributed as JavaScript packages. Although extensions can be written in JavaScript or any language that compiles in JavaScript, we recommend writing extensions in TypeScript, which is used for the JupyterLab core extensions and many popular community extensions.

"However, you will have some problems if you are using the Docker Toolbox, because we still haven't figured out how to enable file sharing," he warns, but assures that he will spend some time on this today with the team to figure out the problem. If unsuccessful, any team member who ends up encountering the same problem will not have to worry about using Docker, instead permitted to use Python on their own machine.

Arash asks Kevin if his question was answered, who states his need for the conversation to be continued offline to address some more specific concerns.

"Great. Does anyone else have anything to add?"

"Can you show us how to duplicate an issue card in GitHub?" Stephanie asks.

Cut off-guard, Arash admits "I'm not sure if I know how..."

I find it amusing how he was able to answer the technical questions with much more ease.

"Okay no worries!" she says nonchalantly.

"I guess just create a new one and copy paste." they both laugh at his attempt to give a definite answer.

Arash then concludes the meeting with a promise that he will reach out to the people who want to continue their discussions offline.

"I'll talk to the rest of you tomorrow!"

The meeting finally adjourns on a light note, with several team members expressing their gratitude and bidding their goodbyes.

Kevin

In the training material, developing a template for the frontend of an Echo application was quite straightforward with HTML and CSS. This was nothing different from the work I'd done previously. Trying to render it on Jupyterlab however was a challenge. Thus, I decided to submit my second pull request for the stand-alone frontend built in Django before I move on. Followed by a 1 on 1 session with Arash on submitting my pull request, I wanted to go over this process again to make sure I understood it. Therefore, I spent some time to let it sink in and trace back each of the steps he went over. This is starting to make sense except I made a mistake.

I wasn't sure if I was supposed to push before submitting a pull request, so I did without pushing my changes. How could that have happened? Well, I submitted with a push I did previously for the backend service. Next day, Arash responded. He asked me to investigate the reason there is a conflict. Of course, the issue becomes apparent at this point. But I needed some time to figure out how to fix it. For some weirdest reason, the Jupyterlab extension could not be installed even after I ran all the necessary commands in the Anaconda base environment. I asked this question on Slack, but no one seemed to have gotten that far yet. Arash tried to help by pointing me back to the instructions on GitHub and it worked even though I had entered the same set of commands before. Why would it work this time?

The next day, I was pretty frustrated, simply because this was taking over a day, and I was clearly getting nowhere. I felt uncomfortable asking the question again on Slack, because I didn't expect anyone to give me any working solutions. But I did anyways. Luckily, someone pointed me to a blog with an example of showing an astronaut picture on Jupyterlab. With this example, I finally was able to see the extension. In this case, my issue was I needed to create a separate environment from the base and install all the prerequisites. Somehow, this was the part not included in the instructions on GitHub.

Chapter 9: Practices and Architecture

Friday, May 8

"If you think good architecture is expensive, try bad architecture."

- Brian Foot, Joseph Yoder

The standup meetings following Monday and Tuesday followed the same flow; each team member reported their progress towards completing the training material as well as any issues they were facing. What stood out as a common enemy for virtually the entire team was installing the necessary tools, environments, and frameworks on their machines. From the mentors' end, Masoud has provided more frontend related learning material and Yousef has started creating a domain model for understanding the system. Arash had also scheduled workshops to cover the teams' major concerns.

Chelsea

Tuesday was incredibly frustrating. I could not install Django regardless of the instructions in the tutorial, so I reached out to Arash and then the group as a whole. I took the better part of eight hours and numerous people's help to finally just get Django installed. It was one of the most demoralizing days I've experience in a really long time. On Wednesday, I worked though the tutorial until I had to install Docker. This was another exercise in futility, though I was joined by a lot of people in my futility. It seems that Docker Desktop was not available for Windows machines unless they are running Windows 10 Pro or Enterprise. I owned the Home edition which meant I was required to either upgrade to Pro or try and sort through Docker Toolbox. Based on the experiences of my colleagues, I broke down and upgraded to Pro, if for no other reason

N. Raeesinejad et al., *The Ignite Project*,
https://doi.org/10.1007/978-981-19-4804-6_9

than to save my dwindling sanity. Everything seemed to go okay with the Docker Desktop install after I upgraded, though I did not attempt a connection before Arash's tutorial for us to go through the backend together.

The first demo meeting has just started. I'm eager to find out how this meeting differs from the previous ones I had observed thus far.

Arash starts off by saying he wants to share something on his iPad. Mohammad guides him on how to connect it to Zoom simultaneously. While waiting for Arash to set up, he urges the rest of the team members to turn on their video and try to get in the habit of dressing appropriately as good practice for future meetings. I notice that all members except for Thomas have heeded his advice and now have their videos turned on. Arash informs everyone that Bassem is not able to attend this meeting. He then proceeds to address some members' technical questions. Finally, he pulls up an agenda on his iPad.

Agenda

- Answering questions
- Housekeeping notes
    - Delete branches after merge
    - Tag Yousef for backend / Docker PRs
    - Tag Masoud for frontend PRs
    - Arash will review all PRs, merged and unmerged, at the end of the day
- Walk through how the demo works
- Planning
    - Move to separate boards for each team
    - Identify who needs more time on training material
        - Remove the training services to clean up the directory
    - System design
        - New services (Arash will create the base and the teams will work on the rest)
    - Pick new stories for those who have finished the training material
        - UI mock-up

"Moving forward, I want to reduce the amount of time spent reviewing to focus on the other projects." Arash reveals. For immediate reviews of backend and frontend, he advises the team to contact Yousef and Masoud. Nevertheless, he will go over all their reviews for the day. Since this week is for training, everyone's work is the same so Arash will show what it will look like rather than going over every single member's tasks. He asks who wants one of their services to be demoed as an example and Mihai volunteers. Arash then initiates a quick verification to see if the acceptance criteria work for the backend on his terminal.

I notice that Arash's voice gets muffled and cuts off from time to time but nobody else says anything. He proceeds to check the frontend.

"This is going to take some time. But the important thing is that your code at the end of the week should run on my system. That's how the demo works."

Finally, some people point out Arash's audio problem in the chat.

Mihai:	Your microphone is cutting out a little bit. Maybe you're covering it by accident?
Michael:	It's fine now.
Paul:	It's not cutting out, but it sounds like it's being moved every once in a while.

Arash reveals that his internet is not good today in response and tries to quickly move on.

"Now let's see if we can activate this JupyterLab environment." Arash declares, "To save time in the next demo sessions, I will run all of these, so make sure that all of your changes are merged by Friday at noon so that I have time to run and get everything ready for the demo." After a couple of minutes of silence, an error pops up on Arash's terminal but he remains cool about it.

"Instead of spending time debugging it, the feature is marked as not passing the demo and is moved back to the backlog so that it can be investigated in the next sprint. This means that on my system, this feature is not complete." With that, he concludes the demo portion of this meeting.

He announces that they are now going to plan the next sprint and go through a process referred to as "cleanup".

A WORD FROM THE MENTORS

In GitHub, you can archive project board cards to declutter your workflow without losing the historical context of a project. During each sprint, we archive our finished tasks to clean our board.

"Someone does the development work, and it gets reviewed and merged. Then, another member tests it and makes sure it works on their system. If it works, it gets moved to Done. I will look at things in the Done section on Friday at noon and prep them for the demo. Today is an exception. We will not go through this approach. Instead, I ask each of you to give me an update. Tell me which feature is yours and if you're comfortable enough for the backend and frontend, we can assign you a new task. Otherwise, you can continue the training material."

Arash prompts Team 11 to be the first team to give their update, calling on each member one-by-one.

Yassin reports that he has one task In progress and Merged for the backend. He agrees to have them archived by Arash. Jerome reports that his ⊙Django REST task is in progress and he feels good about the training material but wants to spend more time on JupyterLab and React. Arash says if he completes training during the next week, he can be assigned to a new task.

"By the way, I have created a project board for each team so that tasks are easier to use." Arash remarks.

Lastly, Yazan says that he also feels good about the backend and Rest, and that he also needs more time for JupyterLab and React. He decided to keep his Frontend portion task in progress and confirms that he has no other tasks.

Arash moves on to A-Team next.

"I think I have the backend sorted..." Chelsea starts off hesitantly, but then quickly corrects herself. "Wait that's a lie, I think I have it *mostly* sorted." She further clarifies that she needs to keep her ⊙Connect with Jupyter task in progress in order to sort out her issues with the backend and finish the frontend.

Stephanie reports that her backend task is in `merged` so Arash archives it with her frontend task which is `in progress`.

Next, Toya says, "I feel pretty good about the backend stuff. I feel good about the frontend and Jupyter as well." He confirms with Arash that he is confident with archiving all his tasks and starting a non-training task.

Arash skips Bassem who is absent, saying they will talk sometime after this meeting.

Michael reports last that he is just getting started with React and will need a day or two to finish his training material.

Finally, Arash calls on Team Rocket.

Paul is the first to report, "I have Django and Rest under control, but I'm still in the middle of learning React." he ponders for a moment and then reveals, "I think I forgot to put my issue on the board." Arash says it's fine and notes that Paul will also work on training in the next sprint.

Peter says he's comfortable with the backend, so Arash archives his associated task for it. He adds that he needs to integrate his ⊙`Standalone frontend` task with Jupyter and Arash immediately moves it to `in progress`.

Thomas also reveals he needs more time to learn React.

"I don't know," Mihai starts off once it's his turn. "It worked on my computer, so I thought it was done." I believe he is referring to the error that Arash discovered on the frontend portion of his service. "My plan was to go through React and read up on authentication for Django."

Arash hesitantly replies with "Uhm...sure", causing Mihai to quickly add, "Or whatever you think."

Arash elaborates, "That might be postponed. Let's go over this with everyone else first and then choose tasks".

"I prefer to have more time learning rather than starting a new task." Mihai confesses and Arash nods in agreement.

Mihai

Prior to the official start of the project, I was working through the training material (primarily Django) whenever I wasn't studying for final exams. The coding itself was not challenging but it took (and will take) some time to become familiar with all the different classes provided by Django and how the different files interact with each other.

I think I learned a lot in the first week, but it was also quite challenging and tiring which I expected. I initially finished the backend, then the frontend. However, I did not create my code in the proper templates and coding environment which led to quite a bit of rework. The Docker/Django workshop held on Thursday was very useful and, in the future, it may be a good idea to start with hands-on tutorials.

While I understand and can explain what the code is doing, I am not at the stage where I can code either the backend or frontend without referring to existing code. I am taking a Udemy course to become more familiar with React and looking through Django documentation to get a better grasp of the code. I plan to start working on a feature next week (even if I'm not really confident with the code). What I learned this week was some Django and React, how and why Docker is used, and a little more about the Git workflow.

Setting up the development environment was by far the biggest challenge as we have not been exposed to this aspect of coding too much. Understanding the overall folder structure was another aspect of the development environment which was challenging until we were briefed on how all the pieces fit together. Another challenge was properly using git on a large project; there are many things to remember such as checking if you're on the right branch, committing often, pushing to the right remote-branch, linking the right issue in a PR, etc.. Lastly, it was challenging to communicate with the team exclusively through slack/zoom.

Next, Thanmayee says, "I still have trouble posting, so that's the only thing I need to fix on my frontend."

After hearing everyone's previous similar updates, I'm surprised when Kevin states that he is done with all the training material and ready for non-training tasks.

"We'll see how it goes," he laughs it off and confirms with Arash that all his tasks are already in `merged`, which he immediately archives.

"Advice for everyone! Add an icon or photo of yourselves in your GitHub profiles so I can find your tasks easier." Arash jokingly pleads.

He then prompts all mentors next for their updates.

"I guess I have two issues on `merged`." Masoud curtly states and Arash archives them as well.

"Yousef is probably very unhappy with me merging this without going through the proper process," he jokes, and Yousef says it's ok, chuckling along. I guess there is no one proper process that works for all software development projects.

Arash asks Masoud if he can take on any tasks for the next sprint, and he replies that he is not sure, having been answering the team's questions all day.

Last, Yousef reports, "I had a task for the domain model, and I should work on the user story in the next sprint." This concludes everyone's updates.

"Ok!" Arash claps his hands. "We have 3 people - Yassin, Toya, Kevin - that definitely need something new to work on for next week and potentially others who will finish training need to start something new in the middle of the week. Now, we are going to do some planning where I will first give you a high-level idea of what we're going to do and then try to pick something that is easy to do that does not involve too many dependencies. The tasks that we discussed last sprint might be too much and I will ask if a task is too much for you and whether we need to change it."

Before moving on to the planning section, Paul asks a question about duplicating or moving tasks in the new team project boards.

"Good question." Arash says and proceeds to show how this is done. He first searches for one of Team Rocket's issues and drags it to `in progress` in the ⊞`Team Rocket` board. Afterwards, he deletes that issue from ⊞`Ignite.`

"I will do this cleanup myself, however, moving forward, you guys should have your tasks and PRs in your own team project boards. I will be using this general board for the mentors' tasks in the future."

He shows a new list on his shared iPad screen titled "System Design" and begins to go over how each of the general components on the list are connected to each other for the team to gain a higher level of understanding of the system they will be building.

- Authentication
- Profile
- Lesson
- Course
- Analytics
- Chat

He starts explaining the workflow involving the first 2 components on the list: Authentication and Profile.

"Let's say we have a service called profile and another service that does authentication for us. How this works is that the user browser will send a request to the authentication server with a username and password and the server will respond with a token. Then, when the user browser asks the profile service to create a profile, the user will also send the authenticator token. The profile service is not going to immediately trust the information coming from the user browser, so it needs to verify whether the request is valid or not. So, it sends the token [that it had received from the user] to the authentication server, stating that someone with this token is claiming to be a particular user and asking if this is correct. If the authentication server verifies the token, it will return additional information such as the user's role. Now that the profile service knows that the request is coming from a legitimate user, it will do the work requested of it."

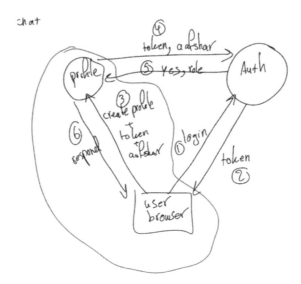

He tells the team not to worry about authentication in the next sprint and only care about the requests for now, pointing to the profile and user browser. This means that their profile service will just blindly trust the user browser for the time being.

"We can follow this workflow for this sprint, until everyone becomes comfortable with how the fronted and backend communicate with each other. If I add authentication to your tasks now, it will only complicate things further."

He then goes over how many frontend and backend their system will have, as well as the relations between them. He starts by drawing the screen from the perspective of the user.

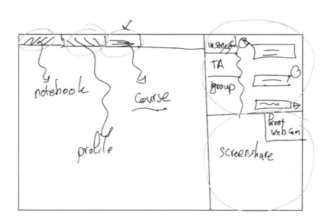

While explaining what the system should look like, he circles the tab in green, pointing each out as a Jupyterlab extensions[29] which they also refer to as the frontend

---

[29] Fundamentally, Jupyterlab is designed as an extensible environment. Jupyterlab extensions can customize or enhance any part of Jupyterlab. They can provide new themes, file viewers and editors, or renderers for rich outputs in notebooks. Extensions can add items to the menu or command palette, keyboard shortcuts, or settings in the settings system. In fact, Jupyterlab in its entirety is a collection of extensions that are no more powerful or privileged than any custom extension.

interchangeably. Each of these extensions works with at least one backend. Arash promises that he will create a base backend and frontend for every new service, to make sure that they are configured, formatted, and linked properly for the team to develop different features on them.

Next, he explains how this is connected to the first two components he discussed earlier.

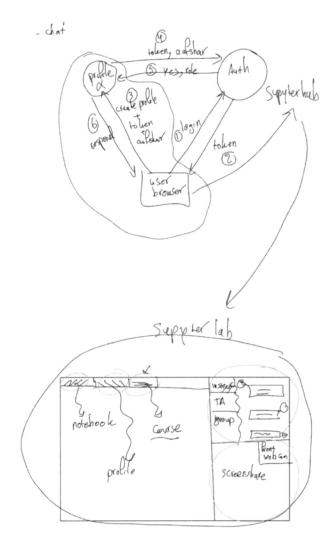

"All the UI tabs are Jupyterlab. The authentication is Jupyterhub. The user browser will send a message to Jupyterhub, which will relay it to the proper extension in Jupyterlab. Working with Jupyterhub is very similar to working with Jupyterlab. However, it has some small issues that I don't want you to worry about right now, which is why we are putting Jupyterhub and authentication aside and only focusing on Jupyterlab for the next sprint. Any questions so far?"

Silence.

"... I don't know if I can't hear anything because I'm on my iPad or if no one actually has any questions."

Mihai immediately replies, "I don't have any questions!"

Arash chuckles, "Awesome, thanks Mihai," and then asks worriedly, "Are we overtime?"

"No, we still have an hour," Mohammad assures, "These meetings are scheduled from 3 to 5."

Relieved, Arash announces that they will now move on to the grooming and planning portion of their meeting, where they will plan and assign non-training tasks for developing independent features that don't involve working with authentication or JupyterHub.

Arash

It is important to provide an appropriate level of detail to the interns at the right time. When interns join an established team, they face a lot of new and potentially unknown concepts: languages, team dynamics, architecture of the system, etc. My experience so far has been that people retain only the information that answers their immediate question/problem. As a result, it is best to release information gradually such that they end up using that information immediately.

Chapter 10: Grooming

Friday, May 8

We don't need an accurate document. We need a shared understanding

- Jeff Patton

"Which feature should we start with?" Arash asks Yousef, who suggests maintaining the labels of each lesson in the system. Nodding, he opens a new issue in ⊟`ignite-aranite` and announces that they will be creating a story for the feature as a team. He notes that since the three members who are finished with their training - Yassin, Toya, and Kevin – are each from a different team, they will need to create at least three stories so that they can start working on them and later be joined by their respective teammates who will finish their own training shortly.

Maintain Lesson Label

He then prompts Mohammad to explain the feature from the point of view of a user and give an example of a lesson label.

"Let's say loops." Mohammad suggests.

As an instructor I want to maintain the labels of a lesson so that the students know what are the main things they will learn in that lesson.

"Now, let's list some acceptance criteria. What do we want to happen?"

"We want to add and remove labels and not have them be duplicated." Yousef explains.

"By the way everyone, this is a discussion, so join in, ask questions, and critique the work. We want everyone to understand and be on the same page."

"In the release planning markdown file, where is this *maintain lesson label*?" Toya asks.

"Good question, let's see." Arash opens the file and scrolls down to the lesson management heading. The first feature is creating and editing a lesson, which, listed as

N. Raeesinejad et al., *The Ignite Project*,
https://doi.org/10.1007/978-981-19-4804-6_10

one of its acceptance criteria, requires for the instructor to be able to fill in some fields, including lesson description, profile link, start and end date and time, lesson title, keywords and prerequisites. I notice there is no mention of labels in this list of fields.

"I guess labels should be keywords instead," Arash notices as well. "Good point, Toya!"

"I don't want to jump ahead," Mihai suddenly says. "But just to get the context, does a lesson model exist in the codebase as it is?"

"No. As part of this story, we should create that."

"Okay."

Arash proceeds to replace the word "label" with "keyword" in the issue description and adds a couple of pre-established AC.

AC:
- Create a keyword
- Edit a keyword
- No duplicate keywords are allowed in a lesson

"Now, we can create these keywords locally per lesson or have a general list of keywords and then have each lesson use them."

"I think we should manage the keywords, so that we have a quantity of created keywords." Yousef suggests

"Okay, so this story isn't really talking about lesson, but just about those keywords." Arash clarifies.

"Yes of course," Yousef agrees. "And we should not be able to remove a keyword if it was used in an earlier lesson." While Arash is adding his suggestion to the AC list, he also asks whether keywords can contain spaces or not.

Arash ponders for a while before saying, "I guess that would be a question for Mohammad."

"Yes, they can have spaces, and should be separated by commas." Mohammad verifies and Arash types as so.

"What else?" Arash prompts.

"Can I jump in there?" Peter asks. "Is it such a bad thing to have duplicate keywords? For example, on D2L, it's not exactly the same thing, but you can't upload two files with the exact same name because it creates a lot of confusion with downloading them, especially when they're differentiated only by a dash and number. Could they be uniquely identified with something like timestamps or instructor name, other than just a keyword?"

"I believe you can think of a keyword as a label for an issue on GitHub," Arash explains. "All the label names are unique, and you can assign multiple labels to an issue. This is the same kind of model, where you have a bunch of keywords and you can assign a subset of them for a lesson. In that sense, the original list of keywords shouldn't have any duplicates but can be used many times for different lessons."

"That makes sense." Peter nods in agreement.

Mohammad adds, "I think having keywords would be useful so we should maintain them, especially when more courses would be offered on the platform and you want to search for a specific class later. Originally, I think what's going to be important is learning outcomes. You basically want to have a description of what you will learn by the end of each lesson."

"Right, so a class could have a keyword or tag – though tag could be a better word here." Arash chuckles. "For example, a tag could be *Python*, so that you can find all courses that cover something to do with Python."

"Absolutely," Mohammad nods. "So then, my question is: Do we want to have a keyword for each lesson or course?"

Arash replies, "I view them as tags which are unique and global to the system for searching purposes. However, those learning outcomes that you mentioned are specific to a course and we don't need to have a global list of all learning outcomes."

"Alright, that makes sense." Upon Mohammad's agreement with his view, Arash replaces the term "keyword" with "tag" in the entire issue.

"Are there any other acceptance criteria that we want to add here?" He takes a sip from his mug while waiting for someone to speak up.

"Is there a limit to the number of tags you can have per course?" Kevin asks.

"Oh, good point," Arash notes. "Each lesson cannot have more than - for now let's say 10 - tags."

"Wait," Yousef suddenly intervenes. "The last point is about assigning tags to a lesson and not about the tag itself, correct?"

Arash agrees and says that they can split their current story into two stories. He opens another issue titled ⓘ**Lesson model** so that tags may be assigned to lessons. It follows that the AC point suggested by Kevin would be for this new story instead.

Kevin also brings up the character limit for each tag and Arash asks what a reasonable limit would be.

"Maybe something like twenty to thirty characters." Toya suggests.

"Sure," he adds it to the AC list. "But make sure it is configurable, meaning that it's not hard-coded. Save it somewhere in the code so that it can just be edited once."

"What about lower case or camel case?" Mihai mentions.

"I prefer to have everything lowercase, but this is a user requirement. What do you think, Mohammad?"

"I think it's okay to have people write whatever they want for tags." Mohammad answers.

Arash then types the words "Python" and "python", asking if they are the same.

"Yeah."

"I think the admin is the one adding the tags." Yousef says.

Arash frowns. "I don't think admin should do that. Again, using GitHub as an example, we as the users are creating the labels."

"Yes, of course, but in this system, if it may be useful if we can control the tags."

"I think it depends on what you want to do with it," Kevin notes. "If you plan on using the tags to filter lessons or courses, then having users create their own tags is essentially useless, because then you will have many different versions of the tag for python."

"So, this has a solution, where a tag can have two fields." Arash proposes.

Tag (key, visible)

He explains that the visible portion is what the user has entered, and the key is the lowercase version of it. This way, duplicates may be identified for searching purposes. He asks the team for their input.

"Who would be adding these tags?" Mihai asks. "Would it be the professor adding them to their lessons?"

"Yes, it would be the professor," Mohammad confirms it.

Arash then points out that if the tags "python" and "PYTHON" are considered duplicates, then a professor cannot write "PYTHON" as it would violate one of their earlier AC stating that no duplicate tags are allowed in a lesson.

Stephanie then suggests, "Could it be that we just have a set number of tags at launch and then instructors can request to have more tags later so that we don't run into this problem?"

"It's great that we're having this discussion and thinking ahead," Mohammad appraises, "but then again, this is just an MVP that will deal with one course. If you think about it from a usability perspective, I can add different tags specific to every lesson in a course, but there shouldn't be different tags that say the same thing for the whole course. I don't want a hundred different versions of the word python to explain one course, you know?"

"That's fair." Arash nods. "So, based on what I've heard, we want to have a localized list of keywords instead of a global list, and all these rules about duplicates are applied to within a course, not the entire system."

"I think so. Let's say that you have a professor from another university who wants to come here. Then they can write their own tags for their own lessons, and we just have to..." Mohammad breaks his train of through to ponder in silence for a few seconds ask, "What would be easier for users and for us to code right now?"

Arash answers immediately that it would be easier to have all the tags as lowercase, but they can change this for the MVP later if needed and Mohammad accepts his response.

"We now have a good number of acceptance criteria which pretty much define our backend." Arash announces while typing in the description. He adds this new criterium for the backend and moves on to discussing the frontend portion of the story.

"We first need a design." Loading up a new tab, he pulls up the UX tool Figma, where he explains he has created a mockup of the extension they will be using. However, after fiddling around with the website for a couple seconds, he switches back to drawing on his iPad instead, realizing that the website lags too many times while using Zoom.

Arash proceeds to present two general views of the instructor's view of lesson tags on his iPad.

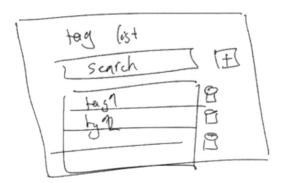 

Mohammad reveals that he prefers the second design, noting that for each lesson, there should be a description and learning outcomes. Arash nods and incorporates this into his drawing.

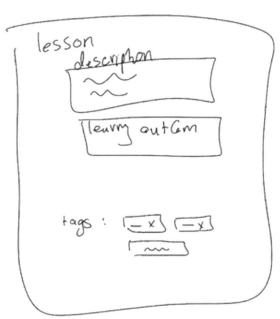

"For the purpose of this task, the UI portion would only include the tags, and the other story about a basic lesson will incorporate the UI portion above tags." He then switches back to share his laptop screen, promising that he will work with Masoud to make the UI designs neater and upload them within the issue.

After verifying with Mohammad that they have covered all points, he assigns the *MVP* label to ⓘ`Maintain lesson tag`.

"Now let's size this story and see whether one teams wants to work on it or if we need to split the work in half for two teams." He posts a URL to Scrumpoker in the chat for everyone to use for entering their estimates.

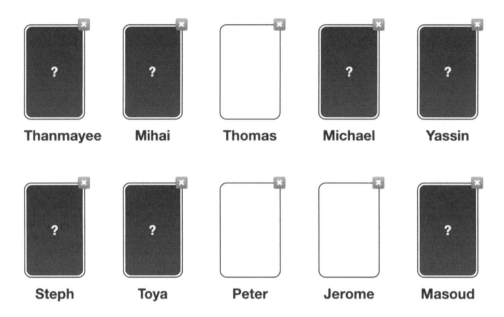

"You estimated the training material as 5, including learning and setting up Docker for the first time and working on both the frontend and backend." Arash reminds. "This story also includes both frontend and backend, but the setup has already been done. You would instead work on things like React and models. Keep in mind that when we are sizing relatively."

A couple of members say that their estimates are not showing up on the screen. Arash tries to refresh his page, but it never loads.

"The problem may be with my internet," he muses. "It's funny that I can talk and show stuff on Zoom, but this site seems unreliable." Immediately afterwards, the website crashes on his tab.

"What?!" Mohammad exclaims in a disappointed tone.

"We'll have to find a more reliable site for next time," Arash sighs. "For now, please type what you think in the chat."

The Zoom chat fills up with the teams' estimates, no-longer anonymous. Arash reminds everyone they can only enter Fibonacci numbers after noticing a "7" typed in the chat. Once all everyone is finished, he scrolls through their submitted estimates. The majority of estimates seems to be "8", followed by "5", "13", and "3". There seems to be quite a variation in what everyone thinks of the size of this story.

"Looks like Michael and Kevin need to fight," he chuckles, seeing their estimates as "13" and "3" respectively.

Kevin is first to justify his view. "For me, I think what mostly needs to be done is creating a model, which is quite simple. In terms of the view, I don't think it does that much other than having an input bar for people to type in their tags and when they enter a comma, then it turns into a tag." He also admits that the reason for which this story may seem simpler to him is because he has already completed the training material, which already covered how to establish a connection between the backend and frontend.

"So, what you are proposing as the solution is to have a text bar which will generate a tag as soon as the user enters a comma." Arash clarifies.

"This is just what I have seen based on other website systems. But to be honest, I don't know what design you guys are looking for at this point."

"Right. Does anyone else want to argue their point of view?"

"I think it's a five," Yazan speaks up. "It doesn't seem like too big of a task and the models shouldn't be too bad, but you still have to decide all the tags and updating them on the frontend side might be a little bit challenging."

Thanmayee mentions that there is also an element of error checking to check for duplicates and verify all other AC.

"Are you arguing that this is an eight?" Arash asks her.

"I originally though it would be an eight, but I don't really have a lot of experience with this stuff." she admits.

"That's alright," Arash assures. "As for the duplicates, we don't want to autocomplete words as they are being typed in since I didn't write it in the acceptance criteria."

> Mihai: Oh, we're not doing autocomplete/showing existing tags?

"Yeah Mihai, in my mind, there was an autocomplete option, but it is not explicitly mentioned. What do you think, Mohammad? As people are typing something, should there be an autocomplete showing existing tags that match?"

"That would be a good feature to have." Mohammad admits and Arash switches back to the ⓘ`Maintain lesson tag` description to update its list of AC. As he is typing, Kevin suddenly asks if he can revise his previous estimate.

"Of course, you can always change your estimates."

Michael: 8

Kevin: 5

Peter: Back to an 8 now with autocomplete

Yassin: 5

"We have 5 and 8 as the majority." Arash declares. "In this situation, I will pick the higher number just to be on the safe side, especially in the early sprints." He then proceeds to assign the corresponding label to the issue - *point-hard (scrum story point 8)* - with a promise that it will give the teams more freedom to explore and time to work on it. He then finally uploads the issue to `ignite-aranite.`

> As an instructor I want to maintain the tags of a lesson so that the students know what are the main things they will learn in that lesson.
>
> AC:
> - Create a tag[tag cannot contain ","]
> - Edit a tag
> - No duplicate tags are allowed in a lesson
> - Remove tags [a tag cannot be removed if it is used by a lesson]
> - A tag should not have more than 20 (configurable) characters
> - All tags will be saved as lowercase
> - As the tag is being typed, the autocomplete should show the matching tags
> - The backend should respond to the typical http verbs (GET, POST, UPDATE) on /api/tags/
> - Frontend mockup: TODO

Noting that there are at least three different sections involved in this story such as search, backend, and autocomplete, he encourages the three members who have completed their training material to work on the story together across different teams as a temporary circumstance, since all other members are still working on their training material. He expresses his hope of this not being the case within the next sprint as everyone would ideally begin working with their appointed teams.

"If this ends up being challenging and the tasks are tightly coupled with each other, we can do the other story for lesson model instead. So, let's refine the description for this story as well, in case we need to work on it."

Thanmayee: We will also need a few hours this week to work on our proposals (due May 15)

"Sure, I'm just refining in case there are dependencies between tasks or those who are done with their training materials need something to do." He assures.

Similar to their previous process, Arash proceeds to fill in the AC for ⓘ `Lesson model` with the teams' input. Certain suggestions involve additional features relevant to the lesson model, but Arash rejects them, in favor of minimizing the scope of this story. Instead, he keeps them as optional criteria that could be either satisfied within the next sprint or later down the road. Other suggestions are listed as "nice-to-have" features to be implemented after deploying the MVP according to Mohammad's judgement. Arash answers some more clarification questions regarding the dependencies between this story and the previous one they planned, since their lesson model would incorporate a list of tags to be managed as well. Mohammad jumps in their discussion from time to time to put things into perspective and offer a high-level view of the live teaching platform that they are building.

After finalizing the AC, they again move on to sizing the story.

"If the story size ends up being too big - for example 13 or above - then we will need to split it into two stories." Arash remarks. The teams enter their estimates in the chat as before.

Kevin:	8	Thomas:	8
Yazan:	8	Jerome:	8
Mihai:	13	Michael:	8
Stephanie:	13	Chelsea:	8-13
Thanmayee:	13	Peter:	13
Toya:	5 or 8	Masoud:	8
Yousef:	8	Yassin:	8
		Paul:	8-13

Arash first prompts member with "5" as their estimates to advocate for their view.

"I just thought that it's pretty straightforward, but I'm not sure." Michael expresses.

Arash hums in understanding. "Once you implement something a couple of times, the pattern becomes straightforward, and you can easily replicate it in other stories. However, it may not be as straightforward for anyone who is new to it."

Toya is next to explain his reasoning. "Looking at it, since the story involves creating, editing, and deleting a lesson, I was thinking the story would be an eight. If we got rid of maybe one or two of those criteria from this story, then I would put it at a five because it's kind of similar to – and Steph and Chelsea can completely disagree with me if I'm wrong – but it's similar to a project we did last semester and our Mozilla training as well."

Arash nods and asks members who entered "8" or "13" to speak up.

"I went up to thirteen because it's way bigger than the tag story." Stephanie says. "There's a lot more components to it, a lot more things that have to work together. We've never done this before. Yes, during tutorials, it's fine, but implementing it is a lot harder than what a lot of us may think. And since this is the first time that we're doing it, I think's it's a thirteen versus anything else."

"I see. Anyone else?"

"I think it would've been a five if we would be working purely on Django." Kevin notes. "But because this one involves working with React and JupyterLab and all that, I think there's quite a few different views to create. That's why I gave it an eight."

"If I remove the edit and delete criteria from this story, will any of you consider reducing your points to eight?" Arash proposes.

After a period of silence in which no one updated their original estimates, he proposes another change. "How about just having create and view lesson with edit and delete but no lesson list and pagination? Does this feel like an eight?"

"Again, it's just the unknown." Stephanie admits, chuckling. "I don't think the edit and delete would change it but taking the list out would definitely make it easier."

"The reason why I am trying to get to the eight point is because if something is thirteen, we should be able to split it into an eight and a five." Arash reveals.

"If we end up deciding on 8 and it takes a longer amount of time or we decide on 13 and it takes a shorter amount of time, what are the repercussions? Stephanie asks.

"None," Arash quickly answers, "We are just trying to make sure that our estimates are correct. After the first two or three sprints, we should all be on the same page and have a good feeling about the size of each page. This is so that when we talk about stories for future sprints, we can have a high level of certainty of whether they'll be done on time or not. For now, these story points don't really matter aside from allowing me to assess where we are."

"I could be wrong, but I think at this point, most of us are gaging this story based on how much of a learning curve it would entail rather than how much work is required to finish it. So, I don't think reducing the number of acceptance criteria any more would make it any easier."

"I see." Arash sight. "In that case, let's make it an eight. We'll see how it goes, and if it becomes too big, that's alright."

He then tells Yassin, Kevin, and Toya that may start working on ⓘ`Maintain lesson tag`, encouraging them to get together after this meeting to split up the story into smaller tasks for them to work on independently and in parallel. If this turns out to not be possible, then two of them can work together on one task. Any other member can join them after their training is done.

"This way, everyone has something to do and if they run out of work, there are more tasks in the backlog that they can start working on. Any questions?"

"When should we complete a task for the keyword story?" Kevin asks.

"Starting on Monday, you will work on one of the tasks that you have created for this story. Each task will have an estimate based on how long you think it would take. Hopefully, it will take roughly the same amount of time."

"So, it's up to us?"

"Yes." Arash nods.

Mihai: If we finish our work early, do we just ask you before we claim a task?

Not noticing the question in the chat, Arash proceeds to conclude the meeting.

"I will create one base service with a frontend and backend for you guys to start adding code to it on Monday. That's it! Bye guys!"

The last thing I see before clicking on the "leave the meeting" button is Mihai awkwardly waving goodbye.

Chapter 11: Feature vs. Component Teams

Monday, May 11

Any organization that designs a system (defined broadly) will produce a design whose structure is a copy of the organization's communication structure.

— Melvin E. Conway

"Let's get started!" Mohammad announces right after I join today's standup meeting with Team 11. *Phew, Right on time.*

It has been a week since these daily meetings have commenced and I am finally getting the hang of following the flow of discussions. Unfortunately, I realize I still don't know the reason for scheduling standup meetings three times a week exclusively for Team 11 - with optional attendance of members from other teams.

Mohammad notifies the team that Arash wants the meeting to start without him. I wonder if it has anything to do with the standup last week during which he urged the teams to take initiative in his absence. Perhaps this is a test to observe how team 11 could direct a standup meeting on their own.

Next, Mohammad asks the other present mentors - Yousef and Masoud - to initiate the meeting, after which Yousef quickly urges Masoud to start instead of him. I'm suddenly reminded of the very first general meeting where I observed Team Rocket trying to work together for the first time in a training simulation, and I find it interestingly amusing how closely the mentors resemble them in awkwardly passing off the responsibility to kick off the meeting.

Sighing, Masoud requests for everyone to wait while he loads the Ignite-Team-11 project board on his screen. In the meantime, Mohammad tries to engage Team 11 in light conversation about their weekend, which he ironically ends up sharing more about than them. A few minutes later, Masoud finally shares his screen and prompts Yazan to start with his update.

"I went through CSS and HTML over the weekend and am now working on connecting JupyterLab with the Frontend. If I have extra time, I want to go over HTML just to understand it a bit better."

After making sure that Yazan has a task on the board, Masoud calls out Jerome to go next. He says he plans to continue working on the JupyterLab section of the training material and reviewing React today, pointing out his task before Masoud can ask.

Lastly, Yassin has been working on the issue ⊙**Create frontend for tags: Autocomplete** for their MVP.

Mihai - who I notice is the only member on call that is not in Team 11 – volunteers to gives his update as well, stating that he has started a Udemy[30] course on React and he wants to be assigned a new task for the user story tomorrow.

"Doesn't your team already have a user story for the frontend?" Masoud questions.

"I don't remember exactly what happened on Friday..." Mihai admits, "There was one person from each group that got assigned to the same user story about the lesson tags, but I don't know if the lesson part got assigned."

Masoud, also expressing that he is not sure, asks Mohammad for clarification.

Mohammad explains that he will have a chat with the MEng group later today regarding the academic merits of the projects that they are working on. "Everyone's writing a proposal right now about which areas they want to focus their work on. We decided to take a feature-based approach. We're not building the site upfront but taking an agile approach to building the platform. My recommendation is that people take part in the agile process but simultaneously focus on one part of it in terms of technology. The PURE students are working full-time, and the MEng students are working part-time. As such, the work of the MEng students should be focused and not spread over broad areas so that they can maximize their learning and efficiency in one particular technology."

It seems this will be made clearer in future meetings. On that note, the standup ends.

---

[30] Udemy is a massive open online course (MOOC) provider aimed at professional adults and students. MOOCs are an online course aimed at unlimited participation and open access via the Web. Whereas regular online course focus more on content, MOOCs focus more on context, making their content more dynamic and relevant.

Mohammad

We started the day with a productive standup this morning. Although this standup was scheduled for Team 11 and optional for others, it was very encouraging to see members from other teams like Mihai joining to present and ask questions. The learning curve is certainly steep but from the current progress and the nature of questions, it seems we are doing very well indeed!

Putting on my educator hat today, I met with both the PURE students at 9:00 am and the MEng students at 12:00 noon. I wanted to communicate to both groups that learning in this project is paramount and encourage everyone to focus on that. I also wanted to highlight the importance of team building which is more challenging with working in a remote environment.

It is evident that our team is progressively getting better at organizing and using tools. Specifically, I think the team is learning to make the most of Slack. For example, some team members like Paul are posting resources there to help others. It is also wonderful to see team members like Yassin and Yazan actively answer questions on Slack. I have high hopes that this will continue to increase which would take some of the loud off our leads, Masoud and Yousef, and will allow them to focus more on other tasks such as development, code review, and design.

A WORD FROM THE MENTORS

A **component team** is a single component and cross-functional team that develops software to be delivered to another team on the project rather than directly to users. This is employed through a "divide and conquer" approach, which makes management easier when one person is assigned to one component (e.g., inventory and accounting). The main advantage of divide and conquer for component-based teams is deep knowledge in specific areas. However, if you regard it as "system thinking", there is only one person in capacity per department. In the case that one component exceeds the others in priority, the maximum capacity for that component would be less than the total available resources. This approach also results in many instances of "handoff", allowing for more bottlenecks, because each developer thinks they did their part and there is lower implication of shared responsibility. Component teams also don't have as much interaction with end-users.

On the other hand, a **feature team** is responsible for end-to-end delivery of working (tested) features. A feature team is a cross-component, cross-functional, and a long-lived team that picks end-to-end customer features one by one from the product backlog and completes them.

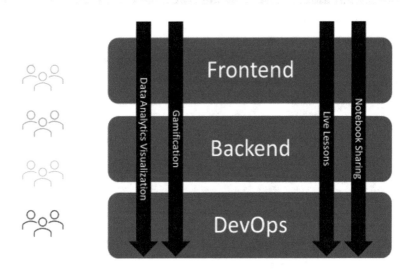

Prior to assigning work to the student interns, there were 4 MEng student teams aka "Agile teams", each of which we wanted to assign components to. The initial plan was to split the students into two teams to handle the front-end and back-end separately. However, we found that structuring teams around the layers of an architecture I.e., decomposing the whole team into a front-end team and back-end team leads to many problems such as:

- ✖ Reduced communication across the front-end and back-end teams
- ✖ A feeling that design by contract is sufficient
- ✖ Ending sprints without a potentially shippable product increment

Consequently, we decided to have larger feature teams by combining the MEng student project teams. A feature team has many advantages:

- ✓ Reduction of waste created by hand-offs
- ✓ Maintained focus on delivering features and delivery of maximum customer value
- ✓ Feeling of ownership and shared responsibility
- ✓ Focus on system productivity instead of individual productivity
- ✓ Simplification of planning

By working on different components of a feature such as frontend, database, backend, testing, etc., students were observed to develop a more well-rounded understanding of how to develop such components. It is important to note, however, that this approach caused some confusion in terms of how the MEng team would be evaluated for the project courses. Although our preferred approach was to create feature teams, a component team may be a good team structure when we want to achieve the following:

- ✓ Build something that will be used by multiple feature teams
- ✓ Reduce the sharing of specialists
- ✓ Reduce the risk of multiple approaches

As it turned out, members of certain Agile teams i.e., feature teams took it upon themselves to further divide their feature work into components, such as the gamification feature:

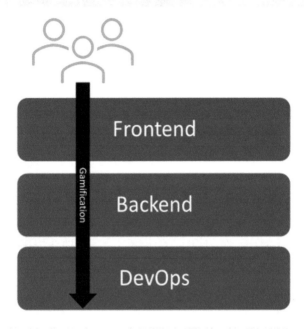

Meanwhile, other teams proceeded with the feature team approach, through which all members collectively worked on all the components of the assigned feature, such as the data analytics visualization:

Arash

Initially, I believed that a component-based approach managed by Agile practices would be the best fit. Using Agile, both myself and the team would get fast feedback loops and the component-based nature would help teams focus on one topic. Moreover, the component-based approach would match what students are used to (working on a defined project and delivering it by a certain deadline). After discussing with Yousef and Mohammad, we decided to take a different approach: allowing all the students to have the option of working on any task, regardless of the initial project assignment. This approach benefited some students in terms of getting exposure to all parts of the system and becoming more informed and involved in the decision-making process for the rest of the project. Moreover, due to some initial linear dependency between projects, this approach provided an opportunity for everyone to work on a task without being blocked by other teams.

However, this approach also had its shortcomings, the most important of which was the confusion it caused for students; initially, I had spoken with each group and decided on the components that they would be working on. As a result, it became confusing for them as to why they are now working on something else. Moreover, this approach made grading a capstone project harder. Nevertheless, students quickly adapted to this new approach and later, once the sequential dependency of projects were resolved, each team started working on their own project.

Overall, this was a good approach, but it could have been better communicated to the students. Moving forward, I would probably start by having a senior developer implement the initial phases and create the structure of the project, and then have each group of students work on their own components. I would then encourage students to review each other's work to get familiarity with the entire code base.

Mohammad

There is no silver bullet with choosing one team structure over the other; it all depends on the nature of the project, timeline, and anticipated workload. It is also important to consider the existing inventory of knowledge, skillsets, interests of the team, especially if they are comprised of student interns. In our case, we found that the feature-based approach provides more freedom by allowing for more customizability. However, one of the traps to watch out for in this approach is that someone really strong in one area of expertise could develop the tendency to implicitly shut down ideas from others, which would make them less inclined to contribute to that particular area.

Section IV. Spikes, Practices, and Guidelines

Chapter 12: Continuous Integration

Tuesday, May 12

"If you want to go fast, go alone. If you want to go far, go together."

—*African Proverb*

8:27 a.m.

I click on the meeting link and wait for it to load on Zoom while I sip my tea. Once the meeting window opens, I notice Yazan is already on the call and mentally applaud him for his punctuality.

The rest of the team members join in a matter of minutes with tired *good mornings* and *hellos*.

At exactly 8:30, Arash says, "I can't share my screen, so we need to wait for Mohammad to come." That explains the lack of pre-meeting chitchat.

"Can I ask something or is it too early for that?" Mihai asks hesitantly.

Arash smiles, "Technically we're one minute in the meeting so go ahead."

Mihai proceeds to seek clarification on where a story card and its divided tasks would exist if they were assigned to two different teams.

Arash answers, "This shouldn't happen, right? Each story should be done by one team. For now, we don't need to be that strict."

"You said that Masoud is doing the frontend for the lesson model..."

He tells Mihai that he can participate in the lesson model story as well, but quickly interjects himself. "Yousef, is this against protocol?"

"It is fine." Yousef chuckles.

"Right. It's not ideal, it shouldn't be like this, but it's okay for now."

I wonder how long the team will operate like this before following said protocols.

Given Mohammad's continued absence, Arash encourages everyone to open their project boards themselves, declaring that they will follow the order of teams from top to bottom.

"Ok, Team One!"

No response.

After a while, some members asked if he meant the first team or Team 11, which is pronounced as "Team One One". Flustered, Arash calls out the latter.

The first member, Yazan, reports that he is getting the JupyterLab extension set up. Jerome follows with his update of having finished the JupyterLab portion of the training material and is planning on reviewing React today. Lastly, Yassin says he is working on creating the frontend for tags. This must be the story created during last Friday's demo meeting.

Verifying my assumption, Arash says, "By the way, I uploaded the UI sketch for the tags story so please take a look at it."

Team Rocket is next.

Paul is still working on React, stating that he wants to gain a solid foundation and finish the frontend by the next standup meeting. Peter is set to finish the training material by tomorrow but admits he couldn't figure out an issue with the frontend and needs help from Arash to troubleshoot, who quickly agrees. Thomas is finished with the frontend and plans on continuing to learn React.

I notice Thanmayee unmuting herself to speak but she gets immediately interjected by Mihai - "Sorry my mic was muted!" - who continues to report that he is working on the backend for ⓘ **Lesson model** and has moved the story to `in progress`. "Does that make sense, or should I have not moved the story?"

"If there is a task in progress, we can leave the actual story in to do." Arash explains as he drags the story back in `To do`. "Once everything is finished, we can move it." Noticing that there are a lot of people working on this story, he also unassigns Mihai from it and instead assigns him to the task he is currently working on for the story.

"Who's next?" I notice Thanmayee has muted herself again, looking hesitantly at the screen.

"I guess I'll go next," Kevin says after a moment of silence, then reports that he is working on the backend for the tags story and reminds Arash of a few questions he had submitted through a pull request.

"I will talk about them after the standup," Arash promises.

Last on the team, Thanmayee says she has finished the training material and is now working on learning React, for which Arash tells her to create a task. She also says she can work on a new task, to which Arash assures he will create a couple of them and update the project board later today.

After another moment of silence, Arash asks, "Who is remaining from Team Rocket?"

"I think we're all done," Peter replies.

He moves on to A-Team.

Michael reports he has his frontend mostly done, but still has something to figure out for the backend. Stephanie has learned more about React and thinks she will finish training today. She also requests to be assigned to new tasks for the week and Arash assures again that they will refine a bunch of tasks shortly.

"I'm pretty much in the same spot," Bassem says next, "I'm just having a little bit of trouble getting the GET and POST to work."

A WORD FROM THE MENTORS

The Hypertext Transfer Protocol (HTTP) is designed to enable communications between clients and servers. It works as a request-response protocol between a client and server. So, HTTP defines a set of request methods to indicate the desired action to be performed for a given resource. Although they can also be nouns, these request methods are sometimes referred to as HTTP verbs. They include GET, HEAD, POST, PUT, DELETE, CONNECT, OPTIONS, TRACE, PATCH.

Here we describe the most important HTTP verbs or methods.

GET:
GET is used to request data from a specified resource. The query string (name/value pairs) is sent in the URL of a GET request.

```
"/test/demo_form.php?name1=value1&name2=value2"
```

POST:
POST is used to send data to a server to create/update a resource. The data sent to the server with POST is stored in the request body of the HTTP request.

```
POST /test/demo_form.php HTTP/1.1
Host: ignite.aranite.com
name1=value1&name2=value2
```

PUT:
PUT is used to send data to a server to create/update a resource. The difference between POST and PUT is that PUT requests are idempotent. That is, calling the same PUT request multiple times will always produce the same result. In contrast, calling a POST request repeatedly have side effects of creating the same resource multiple times.

```
PUT /new.html HTTP/1.1
```

DELETE:
The DELETE method deletes the specified resource.

```
DELETE /file.html HTTP/1.1
```

Chelsea reveals that she is still working on getting her backend to work properly and had spent most of yesterday trying to figure it out.

"Oh yeah, and I will also put up my picture!" she says as an afterthought.

Arash laughs in surprise and thanks her. I don't think he expected anyone to remember his plead from last week to have their profile pictures posted on GitHub to better navigate through task assignees.

Lastly, Toya is working on the layout of ⊕**Maintain lesson tag**. "I got the mock layout and just need to work on the functionalities of the different buttons and get the GET and POST requests up there."

"Same thing I told Yassin," Arash reminds, "The parent story of the tags have a mock-up of the UI now."

"Yes, I saw that." Toya affirms.

This concludes every team member's update.

"Everyone has something assigned to them for now which is great. Let's move on to the mentors." He opens the ▥**Ignite-Mentors** board.

Masoud volunteers to go first. "I am working on the lesson story. Who is working on the frontend part of it?"

Arash replies, "No one but Mihai will be working on the backend."

"Good, I was worried." he sighs in relief. "I will work on the frontend then." Arash clarifies for the team that once he is done, Masoud will use his ⊙**Create learning materials** story for the lessons as a workshop to show them some of the common patterns of creating UIs.

Next, Yousef reports to have been working on the ⊙**Domain model**. He also reveals there is a backend task in `review in progress` on ▥**Ignite-Team-Rocket** for him to review, however he couldn't find it.

"That's because they tagged me in the pull request," Arash reveals and reminds everyone once again to make sure to tag the appropriate mentor for the type of review they are requesting for their PRs.

For his own update, Arash announces that he has created a new template for pull requests.

Resolves #
DoD:
- ☐ Code successfully compiles/runs on my machine
- ☐ New tests for the backend are added
- ☐ All the tests pass when run locally
- ☐ Screenshot for UI change is added

"You may now insert the issue number that the PR resolves at the top. DoD is a reminder of our definition of done for everyone submitting a pull request. Please bear in mind that not all of these criteria are necessarily applicable to every story."

A WORD FROM THE MENTORS

A team's definition of done (often called a DoD) is an agreed-upon set of verifications before any product backlog item is considered complete or "done". A typical DoD takes the form of a checklist which includes the following criteria: the code is well-written, comes with automated tests, has zero-defects, and has been reviewed, and the end-user documentation has been updated and the feature is live on the production servers.

After declaring that they are now in post-standup, Arash spends a couple minutes clarifying some of the AC for the issue, ⓘ**Maintain lesson tag**.

When he reaches the last inquiry from the team regarding what types of tests are expected from them, he opens an existing private repository on his screen.

"You guys don't have access to this repository but I can definitely share how to do this later. We use the same unit testing for functionality, but we use pytests to test the API."[31]After showing the general format of these tests, he promises to send the team an example later.[32]

A WORD FROM THE MENTORS

You can create custom continuous integration (CI) and continuous deployment (CD) workflows directly in your GitHub repository with GitHub Actions. Successful CI means new code changes to an app are regularly built, tested, and merged to a shared repository. It's a solution to the problem of having too many branches of an app in development at once that might conflict with each other. GitHub Actions makes it easy to automate all your software workflows, allowing you to build, test, and deploy your code, as well as manage code reviews, branches, and issue triaging all in one place!

When you commit code to your repository, you can continuously build and test the code to make sure that your commit doesn't introduce errors. Your tests can include code linters (which check style formatting), security checks, code coverage, functional tests, and other custom checks.

You can configure your CI workflow to run when a GitHub event occurs (e.g., when new code is pushed to your repository), on a set schedule or when an external event occurs using the repository dispatch webhook.
GitHub runs your CI tests and provides the results of each test in the pull request, so you can see whether the change in your branch has introduced an error. When all CI tests in a workflow pass, the changes you pushed are ready to be reviewed by a team member or merged to master. When a test fails, one of your changes may have caused the failure.

"So, this is just for testing the backend, right?" Toya asks.

Arash nods. "We create a new test for everything we do in the backend and run all previous tests. The last criterium is for the frontend; Whenever you submit a PR, attach a

---

[31] Pytest is a testing framework which allows for writing test code using Python. Although we can write code to test a database, API, and even UI, a pytest is mainly used in industry to write tests for APIs.
[32] Pytest is a testing framework which allows for writing test code using Python. Although we can write code to test a database, API, and even UI, a pytest is mainly used in industry to write tests for APIs.

screenshot of the UI for the person reviewing the issue to have a rough idea of what it should look like."

Lastly, he reminds the team again that he will be adding new tasks in the backlog for them to assign themselves to whenever they run out of work. He also mentions that he has created a new label for issues called *triaged*.

"When new issues are created, they don't have this label. Once I go through everything and make sure that the story and tasks are good, I will add it. So, if an issue does not have the triaged label, it either means I haven't seen it yet and I don't know what it's about or it is not assigned to anyone and does not yet have a destination on a project board."

Arash then concludes the meeting, and I leave in a chorus of *Thank yous* and *Goodbyes*.

Thanmayee

Peter, Paul, Thomas and I received a new task for researching ways in which we can incorporate video and audio streaming and screen sharing into our applications for lesson management. I have been reading up on different available services, and a few interesting options have come up which can be responsible for all three of the above tasks. As well, Peter, Thomas and I got assigned to a task for implementing searching and filtering lessons based on tags. However, we were not be able to finish this task until the lesson model was completed and merged.

A major challenge with having multiple people assigned to such small tasks has been that everyone seems to be waiting on each other, which I find has been greatly reducing productivity. I hope that for next week we can each be assigned bigger pieces (even if they take more than one sprint to complete) so that we can hack away at a larger problem and figure it out. The way it has been handled so far I find that the majority of my time this week has been just waiting even though I have a clear picture of exactly what needs to be done in my task and I estimate that it would have taken approximately two hours to complete. If one person or one group can be responsible for say all the lesson model, I think it would greatly increase productivity.

Wednesday, May 13

During today's standup meeting with Team 11, Jerome and Yazan reported to have finished their training material. Meanwhile, Yassin will continue working on the frontend portion for the tags, having finished the layout and now working on functionalities. Arash asks him if he has pushed his changes.

Seeing Yassin falter, he says, "General note for everyone: commit often. As I said before, you don't need to submit a pull request but push your changes up at the end of the day for me to review them and give some advice."

He gives the mentors' updates next, having met with Yousef and Mohammad to break down stories into more tasks for teams to pick up once they finish training. Simultaneously, Masoud is continuing to create more resources for the team in the form of learning materials and sample development templates.

"Masoud has sent a poll for his office hours on Slack," Mohammad reminds the team. "and we're going to have a general React tutorial from him so that's exciting."

I open the general Ignite channel on Slack to find that some people have already voted on the poll for Masoud's office hours.

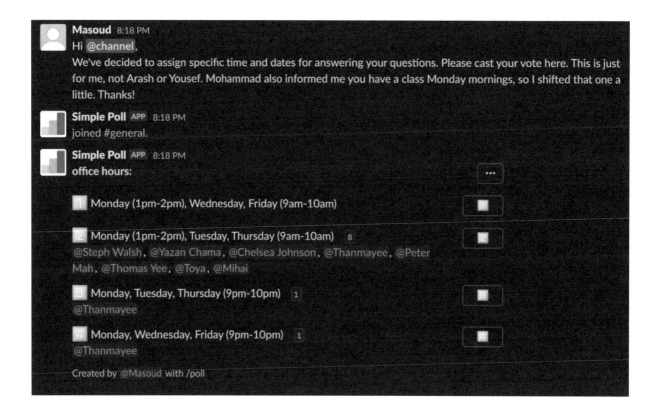

Concluding the standup portion, Arash spends the next couple of minutes explaining a new service that has been added to the Ignite system.

"We now have both a lesson and profile service. The system is going to have three main roles: faculty, admin, and student. Note that they are hard coded inside a class and not in the database. The reason for that is because we're working in a microservice environment and our authentication system and user database are on different microsystems."

He continues to explain that once a request is verified by the authentication service, it will also return the role of the user to the requested service. While the entire database is inside the authentication service, the only piece of information needed by all the other services is the user role potentially along with other credentials such as name or id, depending on the nature of each service.

"I will talk more about this with the people who want to work on the profile service. Jerome and Yazan, let me know once you finish reading through the new tasks and we can discuss the details." They nod in response and the meeting ends.

Mohammad

For the last couple of weeks, we have done well with training and getting started. Since Team 11 – the PURE group – have full-time hours, I'm hoping for them to really come through with problem solving and expecting them to work closely together with Arash for his guidance going forward. We have two major problems right now: document sharing and video sharing. We really don't know how this is done. For everything else we know how to do, we can train our interns. This particular bit is quite innovative, and I believe it's a challenge that the undergraduate interns can definitely brag about as it is not a trivial task.

Chapter 13: Spikes

Thursday, May 14

"Uncertainty is guaranteed to exist at the start of a new project. How a team approaches uncertainty tells me a great deal about whether they are likely to successfully deliver a product users will want."

- *Mike Cohn*

"We're having some pretty nice weather hey?" Mohammad tries to engage in light conversation as soon as he joins today's standup with all the agile teams.

"Is that supposed to be sarcastic?" Bassem retorts and I hear some people snicker. It seems Calgary's always late on the memo when it comes to warmer seasons.

Once all 19 members are present, Arash pulls up the project board on his screen and prompts Team 11 to give their status first. Jerome and Yazan have made progress on creating models for profiles and investigating authentication tokens respectively. Yassin has finished the frontend layout and functionalities for tags and will work on connecting them to the backend.

Arash frowns at his screen. "It says your task is closed."

"I need to create another task for connecting it to the backend. I will do that right now. I submitted a PR yesterday for the frontend and I still need to merge it."

"Don't close the task yourself, we will close it once it's merged," he opens the issue ⊕ `Create frontend for tags: Autocomplete`, staring at the additional assignees in confusion. "Wait, is this your task?"

"Umm, it was created by Kevin and got assigned to me, I think." Yassin answers hesitantly.

After scrolling to the bottom of the issue, Arash clicks his tongue. "It seems Toya has closed this. We'll figure out what happened later."

© The Author(s), under exclusive license to Springer Nature Singapore Pte Ltd. 2023
N. Raeesinejad et al., *The Ignite Project*,
https://doi.org/10.1007/978-981-19-4804-6_13

Yassin

Since I was the only one from my team to finish the training material, I worked with Kevin and Toya this week. Working with them was a very interesting experience for me; Since they come from different backgrounds than me, it was very valuable to learn from their experience. I think working with new people is always nice since there is always a lot to learn from others. Being the first week to work on a real story, we were a little slow to get used to the process. However, as the week went on, we were able to adapt and submitted a pull request by the end.

Paul is first to report from Team Rocket that he has finished the training material and will check to see if can join his team on a task or start a new one by himself.

"Since we're at the end of the Sprint and we can't bring in new tasks, please collaborate with your team members." Arash urges.

Mihai then reports, "I'm working on the lesson model backend. I wrote some test cases yesterday and I want to finish and hopefully bring in the tags. I don't know if the tag model is merged yet, but I want to do that today."

"We can talk about it after Yousef's update." Arash promises.

"I was taking a React course before, but now I'm confident in taking on a task." Thomas says next and notes the issue number of the backend task he wants to start working on in the backlog. Arash proceeds to add the issue card for ⓘ**Search lessons in the backend** to the ⊞**Ignite-Team-Rocket** board for him.

Peter has also finished the training material since the last meeting and is now working on the backend. "Thomas and I had a meeting yesterday and we don't think we can start until the lesson plan that Mihai is working on is complete, which leaves no task for me until the end of the sprint. So, I can either continue learning React or start a new task."

Arash says, "You and Paul are out of work so far for this Sprint. I have a research task that you guys could work on if you're interested. Your team was initially supposed to work on videos, right?"

"Yup."

"Okay, so you may start researching on how to integrate video streaming capabilities into a Django backend. We can define a task for it after the standup."

Kevin gives his update next. "I was working with Yassin and Toya on the frontend yesterday. Hopefully the backend review will be done today so that we can start connecting them together."

"While waiting you can join in on the research task," Arash suggests.

Thanmayee follows up with her status, "Yesterday, I was working with Peter and Thomas on searching for lessons and filtering in the backend. We decided to wait until the lesson model is merged so I can work with them on the research in the meantime."

Arash thanks her and hovers his mouse over ⊙ `Learn React Fundamentals` in the `in progress` column. "This one is merged and done, right?"

"Umm, I guess so. I'm not sure I'm done learning React though." She laughs.

"Learning is always in progress." Arash agrees, chuckling.

Before he can move on to the next team, Paul intervenes with a question, "Do you want us to make a GitHub issue for researching the video?"

"We can discuss this together after the standup," Arash assures.

After opening ⊞ `Ignite-A-Team`, Michael starts off by reporting that he has finished the training material and doesn't have a task, to which Arash promises they will find something for him to do. Stephanie has also finished training and is planning on meeting with Arash to talk about the user service.

"You can also collaborate with Michael on this," Arash suggests.

Bassem says he is in the same boat, but then adds, "I also had a bit of a hiccup yesterday, accidentally deleted my whole repository and had a bunch of other problems..."

"Oh no." Arash says sympathetically.

"But other than that, I'm ready to take on another task today," Bassem chuckles nervously.

"Alright, same thing: work together. We are at the end of the Sprint so we don't want to bring in something completely new that we can't finish by the end of it. That's why I'm asking you to help each other and make sure that the tasks that are on your board actually get done and moved to the end of the board."

Chelsea reports next. "So yesterday I finally finished the backend and I still need to work on the frontend. But…" she sighs dejectedly. "I've either lost the ability to load Django or Django is gone from my computer for some reason, so I have to do some troubleshooting first."

"Ping the channel if you need help." Arash replies.

Lastly, Toya has completed his task for creating the frontend layout for the tags and has also worked with Kevin and Yassin on completing the autocomplete tags. "Today we will be doing some more pair programming to connect it to the backend, and just do some styling to make the frontend look a little bit nicer."

"Awesome! Now that the actual non-training tasks are getting done, we will start going through the proper software engineering practise, meaning that when something is merged, then someone else from the team needs to pick it up, test it, and once that testing passes, move it to Done." He continues to announce that he will remove the training material tasks from the boards without going through those proper steps.

He then hovers his mouse over ⓘ `Create frontend for tags`. "Those of you in A-Team who are out of work and have not worked on this can pick up this merged task and test it and move it to the last column on the board. That's another option if you are out of work."

"How exactly do we test if it's the backend?" Mihai suddenly asks.

"You have the acceptance criteria," Arash answers, opening the parent story ⓘ `Maintain lesson tag`. "The first acceptance criterium here is that the tag shouldn't contain a comma. So, try to add a comma and see if it works. Then try to edit a tag or add duplicate tags. All these acceptance criteria are what you test both for the frontend and the backend. For the frontend, it's just mouse operations like clicking. For the backend, you are using postman or something similar."

Toya chimes in with a follow-up question, "If my task doesn't cover everything in the acceptance criteria for the story, is that a bad thing?"

"What should happen is that you should complete the whole story, so that when all the subtasks are finished, we can then test the whole thing." he pauses to ponder for a couple of seconds before continuing. "Actually, let's do this. Let's test the backend separately and then the whole story together, because testing the frontend on its own doesn't make that much sense." Toya hums in understanding.

The mentors give their updates last, with Yousef continuing to work on reviewing backend tasks and editing the domain model regarding how to implement the application logic for the backend, and Masoud continuing to create learning materials. He announces that he will have it ready for the frontend workshop next week.

"I did a little bit of re-structuring the directories." Arash announces next. "You don't need to worry about it now, do your tasks normally. But just be aware once your changes are merged by the end of today, I am going to merge mine which involves removing all the training material, moving a couple of directories around, and removing all those large files in the history to trim the Git size. We can talk more about the new directory structure later."

He then moves on to creating the research story.

"I have been calling it research, but from now on, let's call it by its proper agile name which is spike."

A WORD FROM THE MENTORS

Sometimes, we don't have enough knowledge about the product or its building process to go forward, so we need to do some exploration. This is known by many names such as prototype, proof of concept, experiment, study, and spike. They are all essentially exploration activities that involve buying information.

Agile teams use the term **spike** to refer to a time-boxed research and exploration activity. Defined originally in Extreme Programming (XP), a spike represents activities such as research, design, investigation, exploration, and prototyping. The purpose is to gain the knowledge necessary to reduce the risk or uncertainty of a technical approach, better understand a requirement, or increase the reliability of a story estimate.

Spikes in a sprint can be used in several ways to:
- familiarize the team with new technology (hardware, software, etc.)
- analyze a problem thoroughly and assist in properly splitting the user story or dividing work among team members
- mitigate future risk

> Spikes primarily come in two forms: technical and functional. Functional spikes are used to analyze overall solution behavior and technical spikes are used to research various approaches in the solution domain.
>
> Some user stories may require both types of spikes. Here's an example:
>
> > "As a consumer, I want to share my notebook with a TA so that I can quickly get feedback."
> >
> > In this case, a team might create both types of spikes:
> > - A technical spike to research how to share a Jupyter Notebook file between two users in JupyterLab
> > - A functional spike to prototype a UX/UI in JupyterLab and get some user feedback
>
> Splitting a spike from a story is may be a good approach as it reduces the uncertainty in the initial story, which will likely make the story take less time to develop.

He opens a new issue and assigns the label titled *point-spike*. "Since we don't know how long research will take, we give it a time limit instead."

> **Video/audio streaming**
>
> AC:
> - How to stream webcam through Django and React
> - How to stream audio through Django and React
> - Screen sharing

"Right now, we're talking about the backend. Let's say we have Django backend application. Instead of having REST API endpoint it needs to have something else so that it can connect to the frontend, receive something from the input, then stream it to the other end with the receiver. How does that work? What kind of libraries and functions do we need? That's pretty much what you need to figure out. Don't code or implement anything, but research and figure out what we need to do. Your research documentations will serve as reference for future discussions on the implementation."

"Does it have to be using the Django framework? Can it use something else?" Peter asks.

"If you use Django, it makes it easier. But, for the purpose of research, let's try to figure out the easiest way of doing this and then we can decide." With that, he removes Django and React from the issue description and uploads the story, encouraging members with no current work to split it into tasks before the end of the sprint.

"One of my students in my MEng class has a lot of experience with AWS and he's going to give a workshop on it for ENSF 609. I wanted to invite Team 11 to attend it as well." Mohammad announces and leaves it up to the team to schedule Masoud's fronted workshop around it. After looking at their poll for office hours on Slack, they agree with having it on Tuesday afternoon.

Yazan

These past couple of weeks have been fairly hectic; It's been a steep learning curve, that is for sure, from learning python, to then learning the Django framework, to then moving to web design via HTML, CSS and React. It sure was a lot to take in! Although I was able to go through the training material, I was still a little hesitant of whether I actually understood the material - As there was a lot to take in- I knew that I just needed to put in more time after the regular work hours and invest my personal time and try to get a low-level understanding of all the material. Personally, when I learn something, I don't feel comfortable with quickly glimpsing over all the material and try to convince myself I understand something when I really don't. I strongly believe that in order to have a good understanding of a topic, you need to get a solid low-level understanding of the topic. Although this might take a bit more time, I believe that in the long run it will be more helpful when it comes to building and designing apps using that language. So, all in all these first couple weeks have been hectic and stuff is a little challenging I believe as the weeks go on and I become more comfortable with these languages - that will take time, patience and coding in them.

Friday, May 15

"I did some research on how to extend the REST framework for authentication and I'm still looking into it." Yazan kicks off Team 11's update for today's demo meeting.

Jerome follows suite. "I'm currently adding functionalities for sending GET requests and getting profiles for each user. I'm not done yet either, still writing tests."

"For your team you potentially have some tasks to work on by mid next week, but Yassin needs some tasks, correct?" Arash asks, noting that Yassin's task for creating the frontend for tags is merged on their board.

"I can join them on the profile tasks." Yassin says.

"Your team seems to be set on this story," Arash observes. "We haven't sized it yet, so let's first quickly talk about what this story entails for the rest of the team."

The $\textcircled{!}$ **User profile** story is about creating a profile for the main three roles: faculty, admin, and student. Arash explains the acceptance criteria regarding how the profiles are created in the backend after a certain event is triggered and limits to how many profiles a certain user can have. For instance, a user can have at most one faculty profile. During the initial setup of the Ignite system, the admin profile will be created manually.

"Does everyone have a basic understanding? If there are no questions, we can go ahead and size it."

"Is it only the team that's involved in the story that does the sizing?" Bassem asks.

"From an agile point of view, it's the team involved with the task. However, based on last week, different people are working on different things and a team's work affects other teams. So, right now, the whole Ignite team should be on the same page and size together."

After a couple of minutes of typing their estimates in the chat, Arash announces a unanimous 5 as the size. He moves onto Team Rocket next.

Mihai gives a rundown of the status of their current tasks. "The tasks for creating the backend for tags and the lesson model were merged from yesterday's big merge. So, I made a new task that is now in progress. I'm going to put the lesson tags and models together and finished the testing for it. We're running into some issues with validating the number of tags and there isn't a good easy Django method to validate them. We tried to research it, but we haven't figure it out yet."

"Spend at most one more hour on it," Arash tells him. "If you still can't figure it out by then, we can just assume that the information will be verified by the frontend. Let's not drag it on for too long."

The rest of their tasks are either blocked until they finish the backend or spikes for researching screen sharing and streaming webcam video and audio.

"From this, I gather that your team will spend at least another day or two researching and then will finish the tag and lesson backend. So, you will need some more tasks for the rest of the week, which we will talk about later."

He moves on to A-Team's board and immediately notices a couple of tasks remaining from the training material. Saying that he assumes they are done by now, Arash archives those tasks and then asks for their update on ⊕`Maintain lesson tag`.

"The frontend is almost done, and we completed most of the acceptance criteria. Since the lesson backend is finished, we should be able to complete it." Toya reports and asks for clarification regarding one of the acceptance criteria for removing tags. Arash quickly explains to him the difference between removing tags in the frontend versus the backend.

"So, the backend is finished and will be tested. You said that there's still some work left for the lesson to be connected and tested. This shouldn't take too long, so your team will need something for the next sprint."

He then asks for the status of ⊕`User service`, which Stephanie, Michael, and Bassem have been working on together and are already creating tests for reviewing it.

"One question we have is how much testing do we need to do? Do we need to do anything more than the pytest?" Stephanie asks.

Arash answers to just check the main four verbs: GET, POST, UPDATE, DELETE. "If you're creating something Inside each of those, make sure it ends up in the database."

"In that case, we're done, and we submitted the PR just a few hours ago."

"This is only for the backend," Arash reminds. "We need to also create a corresponding frontend. Once this is reviewed and merged someone needs to test it. since it's already done, instead of asking the whole team, I'm just going to ask the four of you. What was the size of this story?

"I was going to say three, but Steph wrote a lot more code than me." Bassem says.

"I'd say it was a three," Stephanie agrees. "The issue was just figuring out how to test, but the code wasn't too bad."

Arash nods. "We can have this as a point of reference for building a basic backend and use it to size other stories."

He then moves all their tasks from `Merged` to `Done`, reminding that anything involving the UI should be tested thoroughly and to not include PRs in project boards since they are already linked to their respective issues.

Lastly, he asks the mentors for their status. Yousef has been working on the domain model and doing some other work on the backend architecture, while Masoud reveals to have spent a fair share amount of his time answering questions and is still in the progress of creating learning materials.

"Hopefully this will be resolved this week, with our dedicated office hours."

Before Arash can give his update, Bassem declares he has a question regarding the reviewing process. "I understand that we assign a reviewer, either Yousef or Masoud to review our pull requests. After that, do we have to give it to another team member to review our code again? How does that work exactly?"

Arash explains that Masoud, Yousef, and himself oversee the code review. "We see if it has proper structure. Does it make sense? Are there any obvious problems we can flag? Is it in line with the guidelines? These are all things we can understand and comment on using the human eye. We also run tests for the backend and see if the code compiles. Once your PR is accepted and merged, then the actual acceptance criteria need to be verified. If you go with traditional software engineering, we have a developer role and a QA role for this second level of testing."

"In the case of our team, we wrote our own tests for our story. Does that cover that second portion?"

Arash shakes his head. "Regardless of how thorough you are, the person who writes the code always makes mistakes. For example, someone else might run your code on their machine and point out that you missed a comment about configuration."

"I see." Bassem hums in understanding.

"The final level of testing is what I do during the demo which is now. Given it is done, merged, and verified by someone else, we demo to the stakeholders and see whether it matches what they expected."

Arash points out that since none of the stories are finished, he can't really demo them today and will instead demo their new directory structure.

"I talked enough about cleaning the git history on Slack. The main purpose was to move things around, so that we don't have any more training material and that the actual architecture of our service is correct."

He spends the next couple of minutes explaining the content of each directory. Django applications are logically separated, despite having the same database and port, to allow for better packaging and code management. He also explains how to run JupyterHub and JupyterLab as well as the steps involved in their authentication mechanism.

"So, we cleaned up our boards and finished the demo. Next, we will plan what we will work on in the next sprint."

Chapter 14: Personas

Friday, May 15

"Writing software that fully meets its specifications is like walking on water. For each, the former is easy if the latter is frozen and near impossible if fluid."

— George Box

Arash navigates to the *issues* in the main repository and clicks on *Milestones*.

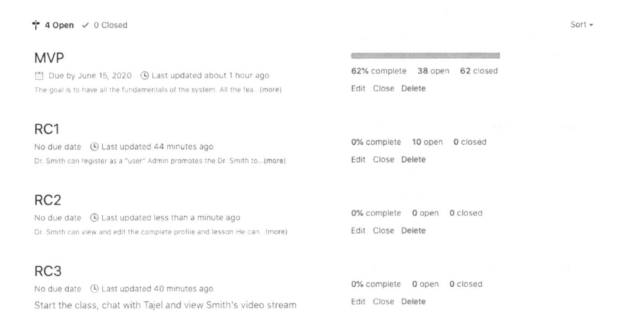

"We have our big picture, the MVP, that we want to have by mid-June. But that is quite far away so we are going to have release candidates, which are essentially a small set of features that are presentable to the stakeholders every sprint and a half. I am using characters from PhD comics as personas for all the different roles in our system. Dr. Smith is a professor, Cecilia is a graduate student and TA and Tajel is another student."

© The Author(s), under exclusive license to Springer Nature Singapore Pte Ltd. 2023
N. Raeesinejad et al., *The Ignite Project*,
https://doi.org/10.1007/978-981-19-4804-6_14

A WORD FROM THE MENTORS

An effective software must always be designed for a specific end-user.

A **persona** defines an archetypical user of a system, an example of the kind of person who would interact with it. A **negative persona**, on the other hand, is someone that you are not designing for. A **primary persona** is someone who must be satisfied but who cannot be satisfied by a user interface that is designed for another persona.

You can use the following techniques for writing effective personas:
- Don't "make up" personas! Instead, discover them as a by-product of your requirements investigation process.
- Write specific personas: you will have a much greater degree of success designing for a single person. The "generic user" will bend and stretch to meet the moment, but your true goal should be to develop software which bends and stretches. Your personas should "wiggle" under the pressure of development.
- Determine the persona's goals to better understand your system requirements and constraints.
- Try identifying some negative personas.
- If you identify more than three primary personas your scope is likely too large.
- Your number of personas must be finite, that is, your goal is to narrow down the people that you are designing the system for.

Personas are incredibly useful when you don't have easy access to real users because they help guide your functionality and design decisions. Questions like "How would Dr. Smith use this feature?" or "Would Cecilia even be interested in this?" help better identify user roles and write stories.

Personas in the Ignite Project:
The personas for our project come from Piled Higher and Deeper, otherwise known as PhD Comics, which is a newspaper and webcomic strip following the lives of several grad students.[33] First published in 1997 by Jorge Cham who was a grad student himself at Stanford University. The main characters used as personas in our release candidates are Cecilia, Tajel, and Dr. Smith!

---

[33] http://www.phdcomics.com

Arash opens the first release candidate.

RC1

- Dr. Smith can register as a "user"
- Admin promotes Dr. Smith to "faculty"
- Dr. Smith updates his profile
- Dr. Smith can create new lessons and view them
- Dr. Smith can publish some of the lessons

Spike: video/audio streaming and screen-sharing

"These are all what we want to have at the end of this release candidate. We also want to do a spike for video and audio streaming and screen-sharing to figure out what we need to do for the next release candidate."

He then navigates to a new project board called ⊞**Ignite-stories**, which serves to provide a high-level overview of the status for all stories in each release candidate. After filtering them with the *milestone:RC1* tag, he points out all stories that need to be finished by the first milestone, most of which are already `In progress`. The remaining stories are in the `Backlog`, yet to be sized.

Arash opens a new issue called ⊙ **Authentication with JupyterHub - first user authenticator.**

"JupyterHub by default works on the username and password that you have used to login to the system, but it is not practical and usable for a couple of different reasons. Also, Google OAuth is going to cause us a lot of problems that we don't want to deal with at the moment. So, we're going use a new plugin called first use authenticator for JupyterHub. How it works is that when you install it, the first time that you enter a username and password, it's going to register. The next time you enter the same credentials, it will login. This makes the distinction between registration and login very easy."

He then goes over all the acceptance criteria for the story regarding how the profile will be created once a user is registered into the system.

"In order to launch JupyterHub, do you need to have an account already?" Paul asks.

Arash nods. "If you want to use JupyterHub, you need to define a username in your system. For example, if you want to login as Cecilia, you must have a username for Cecilia already in the system. A cleaner way is to launch JupyterHub in docker. but given our experience with docker it may cause some trouble with a lot of you and slow down the process."

"Is this for the backend or the frontend?" Mihai asks.

"The part about how to add usernames only involved configuration and documentation. For creating users, we need to figure out how to get JupyterHub to send POST requests through a post registration hook. The last part for showing a warning banner for an empty profile is just UI work. The actual work isn't much; We just need to figure out how to make it all work together."

He then prompts the team to type their estimates and after a couple of seconds, the chat fills up with "8" and "13".

Toya is the first to defend his estimate. "It's more of a gut feeling. Your explanation just sounds like a thirteen."

"I agree." Peter says.

"Let's break it down piece by piece." Arash suggests. "How hard do you think it is to figure out the JupyterHub hook?"

"That's the part that I don't know." Kevin admits.

"In that case, I will take care of that part; I will figure out how to call the hook. Putting that aside, what are your estimations for the rest of the story?"

"After that, I would say closer to a three or a five." Yazan says and Kevin voices his agreement.

"Then we can split this story into two parts." Arash posts a new comment under the issue.

```
Spike: find the JupyterHub hook
The rest of the AC.
```

Mihai then asks about the format of the documentation, confused about who are the actual readers of the instructions.

"This is just for development purposes. It will not be used in production." Arash clarifies.

"Oh okay, that makes sense."

Arash explains further that once everything works, the team will use Google OAuth for new usernames and emails that they haven't seen before. Then, they will run JupyterLab inside Docker. and create usernames, after which everything else will become automatic. The spike for finding the JupyterHub hook will be delegated to either Arash or one of the mentors.

He then opens the next story in the backlog: ⊕User dashboards.

"When a new user is registered in the system, their profile and account are created. The next step in the workflow is to show the dashboard to the user. This is where a bunch of information is shown, which we will take care of later. Right now, we want to make sure that every type of user can see their own dashboard. This will be an empty dashboard with the name of person who has logged in."

Kevin asks if this story involves a backend model or just more frontend work by pulling data from other models. Arash confirms it is the latter, involving roughly 90% frontend work.

"So, there's no customization option? Will every student and every instructor see the same dashboard?"

"Yes, each role has the same dashboard, so we will have three frontends filled with different data."

The majority of the team type in a "5" for their estimates and Arash quickly moves on to the next story: ⊕View and promote users.

"This is why we have separate dashboards. In the admin dashboard, we need to able to see a new functionality, which is an icon that directs to another page and allows them to view all users in the system. This list view has the name, email, and role of each user and the admin has the option to promote a student to faculty. Like the previous story, most of the backend is already there; We have the user profile, and we can get all the information from the backend. We just need to customize the admin dashboard further."

After a couple of seconds, the team unanimously sizes the story as "5" in the chat. However, Arash voices his disagreement with the size, believing it should be smaller.

Paul volunteers to defend their estimate first. "I recognize that it is relatively straightforward if I was already familiar with the software. Personally, I'm still pretty new to react and just have a basic understanding of the frontend, but I don't' have much experience yet under my belt. That is why I see it as closer to a five."

Paul

As part of training, we were expected to implement an echo service using Django REST API for the backend, React/Typescript for the frontend, JupyterLab extensions, Docker, and Git.

To learn Django, I have completed most of a video tutorial on YouTube and worked through some of the Mozilla tutorial provided in class. However, it was incredibly large and wasn't as comprehensive as I would have liked. I felt that the video tutorial had given me a good enough base understanding that I could begin working on the REST API. Initially, I felt pretty lost because I had no idea what a REST API was or what the training materials were even asking for. However, going through another online tutorial, this time on Medium, was an incredible help. Using this resource, I was able to implement the REST API with little issue. I was able to check Arash's example when I got stuck. I also began watching a Udemy course on React.

The learning curve required for this simple application has created quite a workload. It would have been a huge help, seeing as so many new resources were expected to be learned, if a step-by-step framework was established to tell us what needs to be learned, what order should it be learned in, and what (comprehensive and project relevant) resources to use in order to understand the requirements in a relatively short amount of time. Many of us got stumped on one part and skipped to another. Suddenly, everybody was working on different things, and we couldn't assist each other as well. Providing a strong set of resources will ensure that each project member has the same approximate background. When people learn individually, dramatically different resources are used, and team understanding is much more varied. When so much needs to be learned in such little time, I found that the provided resources were insufficient. Software is most quickly learned through comprehensive examples, not pages describing theory.

"I just have a question," Bassem says. "How does promoting someone work exactly?"

"When you promote a person, their role changes. For example, when you query JupyterHub, it will say that this user is now faculty." Arash answers.

"That makes sense, thanks."

"Things like creating a lesson, showing a list view, and going into edit mode are part of a general pattern that happens frequently." he continues to explain. "Once Masoud goes over the tutorial that he is currently preparing, this should become very easy; Eventually, when I tell you to create an endpoint that does so and so, you will immediately understand what to do. This common pattern is the reason why I would give this story a three, but we can leave the size as a five for now."

Kevin asks if he can add a comment. "I've been working with Yassin and Toya on the frontend for tags. For me, the experience was quite troublesome, because right now, we're bound by typescript which we're not quite familiar with. There are also specific coding styles and practices that we are expected to follow, but I'm not clear on what they are. When I search up about react, there's different coding styles, like using purely JavaScript or react frameworks. I'm finding it difficult to implement anything at this point."

"I see," Arash nods. "To answer your first concern, I'm not sure why you are bound by typescript since it should be irrelevant, but the rest of your points are completely valid. I'll let Masoud answer that part."

"Right," Masoud coughs. "There is a learning guide in the repository if you want to check that out. There are a couple of guidelines for react and python. As you said, there are many ways to do something in react and I think pretty much in any programming language. We want to be consistent enough so it's easier to review and maintain your code. So, check these guidelines out. If you see something missing in the guidelines, definitely submit a PR or let us know directly."

Arash adds that it is on his to do list to use GitHub actions to run linting on every commit for both the frontend and backend. He moves on to the last story: ⓘ `Improve JupyterHub Dev Workflow`.

"I'm not sure how many of you have used the watch mode of JupyterLab, but when you start JupyterLab and run your extension in the JLPM watch mode and save your changes, then both your code and JupyterLab compile automatically. These happen incrementally, meaning that only the new stuff gets compiled. Therefore, you have a very fast turnaround; You lose something and then you get the result immediately, within 30 to 40 seconds. JupyterHub, on the other hand, when you enter the run command for it, then it launches JupyterLab itself. This means that by default, we cannot pass that watch mode. Effectively, when you are working on the frontend of JupyterHub, you will make an edit, save it, and then you will need to go to your terminal and recompile everything which will take some time and become very frustrating. Therefore, one of the first things we need to do is figure out how to configure JupyterHub to run in the watch mode so that we can get the same incremental build as in JupyterLab."

"Wouldn't this be a spike?" Toya questions.

"You're right," Arash notices. "Let's not size it." He moves all the stories they have discussed from the `Backlog` to `Groomed`.

The only thing left to address is assigning teams to the stories, some of which are frontend-heavy while others are more focused on the backend. Arash encourages everyone to think about what kind of tasks match their interests and abilities. Going forward, the teams will assign themselves to tasks that they prefer instead of Arash.

"For the A-Team, I know that we're kind of split," Stephanie says. "Some of us like frontend and others like backend."

Peter: **Team Rocket is interested in "view and promote users" and "user dashboard".**

"Seems like Team Rocket is mostly interested in frontend work and A-Team is split. Let me see how I can break things down so that you can work on you like. Please message me directly on what you would like to work on, and I will try to design the stories accordingly. By Tuesday, I will gather everyone's feedback and let you guys know what you can work on."

Toya

This week, I had teamed up with Kevin and Yassin to work on maintaining lesson tags while the others continued training. We used paired programming which helped with learning but waiting for other tasks to be completed delayed completion of this story. Although I understand Dr. Moshirpour's advice to be proficient in either backend or frontend, I still hope to be proficient in both, as my goal after this project is to pursue full stack development career. Working on the frontend tends to be easier for me since you can see the results of the code on the screen, however I like backend more. It seems like more people prefer frontend on this project, which may cause conflict in assigning stories; Most stories to date have also been frontend intensive, with not as much work to be done on the backend.

Arash asks if anyone has any final comments, to which Paul says there is something he'd like to bring up. "We, Team Rocket, are going to be working on a bunch of spikes for audio and video streaming and screen sharing capabilities. But right now, there doesn't seem to be a solid definition of what is required. Sure, there are many things that can stream videos, but we don't know how big of audience is expected and what kind of latencies are allowed, or whether we want to pursue free or paid options. I guess we don't have enough information to really decide on anything or give a good recommendation because we don't know the qualifications that we are trying to meet."

"First things first, let's split this into two ways we can approach this." Arash explains. "The first way is using premade services. They do everything like hosting, and you pay a subscription or one-time fee. Another approach is to assume that no such service exists and build everything from scratch. I want your results for both these approaches. Regarding the size of the audience, it will be people in Calgary with varying internet qualities which the video quality will have to match. In terms of size, I think a couple hundred. Am I correct in assuming this, Mohammad?"

"Yeah, I think that's a safe number. I would even say to bump it up to a thousand, especially if you want to address different schools at the same time." Mohammad scratches his head. "But that is definitely a more high-end way to think about it and a couple hundred may be more realistic." he ponders for another second. "But of course, the bigger the bandwidth we have, the better."

Arash nods. "At the end of the day, this is a business decision which we don't need to make right now. It is a trade-off between the cost and the quality of the service. For now, you can exclude the very high-end services. A few hundred to a thousand users in one region are what we are aiming for and you can look at existing libraries that we can implement ourselves."

Paul thanks him for the extra direction, promising they will let him know if they have any further inquiries along the way.

"The four of us have come up with some solutions," Peter mentions. "When would you like us to give a formal recommendation?"

"Spikes are usually timeboxed, meaning that by end of the preestablished time period, you present what you have. So far, you have worked on this for almost two days, which means that we can discuss the result by the end of Tuesday."

With no further questions from the team, Arash wishes everyone a happy long weekend and the meeting adjourns.

Section V. MVP

Chapter 15: Intern Support

Friday, May 22

"Tell me and I forget. Show me and I remember. Involve me and I understand."

-   Chinese Proverb

Over the course of this past week, the agile teams started to separate and work on the stories to which they were assigned by Arash based on their current interests. Many team members opted to work in parallel through pair programming using the VS Code Live Share extension. While some used this method as a means to work around not being able to use Docker on their own Windows machines, others aimed to benefit from using their collective knowledge towards completing the same tasks. They also started taking more initiative towards modifying stories and adding their own acceptance criteria with permission from Arash during the standup meetings. On the side, most continued learning more React through online resources.

Peter

Since most of us working on the project are still learning the React library, myself and the other group member assigned to the task decided to do pair programming using the VS Code Live Share extension. This was my first time doing pair programming so I had to get the VS Code extension set up and figure out how to use it.

My first experience with pair programming was useful at times and less useful at other times. I found it was quite helpful to discuss the plan of attack with my programming partner before implementing the solution. This forced me to construct complete comprehensible plans before prematurely jumping into the code as I too often do.

Another advantage I experienced pair programming was the ability to get instant feedback on the code I was writing thanks to the second set of eyes.

What I did not like about pair programming was that I often found myself getting distracted by the presence of another person. This may have resulted in me taking longer than I normally would have taken if I were to work independently. But this was only my first experience pair programming so maybe I just need to get used to the workflow.

Meanwhile, Team 11 continued working on more risky and problem-solving tasks in close collaboration with Arash, such as authentication with JupyterHub and documentation and video sharing.

Yazan

So far in my coding career I would learn something in class and then I was expected to apply it in a lab or an assignment, but now I was expected to learn something from scratch and be able to implement it. That meant I had to do a lot of research, watch a couple videos on how it worked before I was able to write the code for it. So, this was something new, and I realized that as a developer, we won't always have someone teach us something, we sometimes have to go out on our own, do some research, and be able to apply it. Although it was a little difficult figuring out where to start, I had some mentoring from my team that set me on the right path I should be taking and was able to complete my task. This goes to show the power of having the support of your team, when working on a project of this magnitude it isn't everyone doing their own thing, everything has to be cohesive and we all have to work together to achieve the end result. Because at the end of the day it doesn't matter if you were the only one who completed the task, you have to work together to ensure that everyone on the team has done so as well and this will ultimately play a big role in ensuring the success of the project.

Mohammad announced to everyone this week that Friday standup meetings have been discontinued. While he didn't explicitly state the reason, I believe it is due to the time constraints imposed by one-week sprints and approaching milestone deadlines. This week's demo meeting bore the weight of this decision, as more time was spent on the standup portion to cover the entire team's individual updates. Unfortunately, there were no done stories to be demoed for this week due to some members continuing to train on the side and troubleshooting any lingering setup issues with their machines. Instead, today's meeting mostly consisted of discussions relating to the UI. At the moment, the team is using only one extension for the front panel, which is where all their frontend work will go. Arash assures that if more extensions will be required down the road, they will add them. However, for now, everyone must follow the same standard, which is especially crucial for working on further stories with shared dependencies.

"Since there are some tasks that are kind of done and need to be merged back for everyone to be on the same page, I suggest we don't bring in any new stories for this sprint. Rather, let's make sure all teams will work nicely together. I'm not going to add any more pressure, so just work on your existing stories. Hopefully we will have some awesome demos by next week!"

Thanmayee suddenly says that she and Kevin don't really have anything to do since they have already finished implementing the frontend for the dashboard.

After browsing through the stories in the backlog, Arash introduces a new story for faculty members to create lessons on the dashboard, explaining briefly detailing each of its acceptance criteria. Since none of the previous features of the faculty dashboard have been created yet, Arash urges Thanmayee and Kevin to create a button in the dashboard as a placeholder which will direct the faculty user to create a course, justifying that he doesn't want their work to conflict with that of other teams.

"Alright, let's size this story."

"Look at us all choosing lower numbers now." Mohammad says proudly as the chat reveals a unanimous size estimate of 3.

"By the end of this sprint, we will have our first release candidate that we can show Mohammad," Arash says excitedly.

With no more questions or comments to add from the rest of the team, he concludes the meeting, wishing everyone a good weekend.

Masoud

On Tuesday, I held a frontend workshop covering how to build a React application and connect all components together. This would then be applied by the team towards building one segment of the project: A JupyterLab extension for lessons. The main goal is to provide the team with a clear picture of how some services should look as well as help them gain a better understanding of React. Although I did not recommend it to them, the approach I used was to first build an application solely with React and then convert it to Typescript and embed in it a JupyterLab extension. I warned members prior to the workshop that they could code along but I wouldn't wait for everyone to finish; This was so that I could only answer general questions and leave specific follow up questions for the end in order to finish the workshop on time.

The idea of integrating full workshops in training is controversial and not the most viable option due to the required time commitment and effort. It is more accurate to call what the team went through as a "working example" rather than workshop. The working example in training for React was helpful, informative and recorded but it had challenges involved. One of the biggest challenges of using examples for training in UI is that people have different design choices. In these cases, the use of code snippets and templates are very helpful towards establishing structure and unifying the design choices of everyone on the team. It would have probably been more effective to incorporate templates ahead of time during training week. If I had the opportunity to host another workshop, it would most likely cover CSS styling, however only this first one was completely necessary for the team.

Arash

I think the workshops were a great help to students. Personally, I prefer students to work on a problem for a week before offering a workshop, since they would learn the contents better and, based on the questions I get, I can design a suitable workshop for their needs. I also prefer workshops to represent actual work (e.g., implement a real feature of the system instead of giving a generic Django or React workshop).

Bassem

We had a workshop on React with Masoud this week. I understood most of the material, but the pace of the workshop was too fast for me to even keep up, we should have split it into two workshops. I hope we have more workshops coming, they have been very helpful. The resources that Mohammad has posted on d2l around React and Django have proved useful to the work I am currently working on. I wish there were more tutorials.

Kevin

I hate to admit, but I was overwhelmed. The combination of JavaScript and TypeScript together was confusing and make it difficult to find relevant resources. Thanks to Masoud, he changed this for me over his workshop. During the workshop, he first develops the app in regular JavaScript + React, and then he converts the code into TypeScript. This was particularly helpful because, in my experience, many resources I find via Google are in the first type. Therefore, it becomes extremely important for us to be able to convert the useful code into TypeScript for the purpose of this project.

Stephanie

Yesterday we had frontend/React training with Masoud. While it was very fast, I think I did pick up on some things while he went through it. I recognize that when I start to do it on my own, I will definitely need to go through the recording, but it was good to know I did learn something from the training materials to be able to follow along. One thing though is that in comparison to the backend training with Arash, we didn't code along. I did find my attention span leaving me after a while. Three hours when you're not super engaged with the content is a lot and might be too much all at once. I'm very happy it was recorded to be able to go back and see it all again.

A WORD FROM THE MENTORS

Student interns often face confusion regarding the right thing to do even when there are no technical issues and their tasks are completed. Learning material for interns require a lot of feedback and modelling. Although potential delays in production and processing can be anticipated, it greatly benefits the company and product for everyone to be on the same page.

The training material and support provided by the Ignite mentors encompassed the following elements:

- Just-In-Time (JIT) video lectures, documentation, and workshops which allowed for students to acquire knowledge relevant to their current tasks, covering concepts such as React, Django, AWS, and the Scrum process. Mentors also created working examples of tricky projects which were delivered through a DevShop style of teaching.
- Constant communication between the students and mentors through office hours and pair programming opportunities.
- Complete models I.e., a framework where all components work together (files, api, connection to backend, etc.). By providing templates already containing pages and classes, students were given the chance to work around an existing framework.
- Evaluation and feedback through code reviews and questions.
- Early anticipation of where the team may experience some challenges and quick identification of areas of difficulty.

There *is* such a thing as overloading with information, so a much more powerful learning experience would be to let people work by themselves. You don't want their mistakes to be too costly where there are huge delays, but you need to let them get stuck somewhere and figure out how to get out with some guidance.

Certainly, the Ignite mentors observed some issues through the approaches outlined above:

- → Students were not sure if what they did was correct after workshops so there was missing feedback.
- → There was a lot of miscommunication in the frontend such as selecting a CSS framework to be used by the entire team. This messed up the progress of students regarding what practices to follow.
- → There weren't many templates made due to time constraints, which resulted in inconsistent code and refactoring that could have been avoided.
- → Since the mentors themselves formed a cross-functional team with varying expertise in frontend, backend, and JupyterLab, it became more challenging for one mentor to answer questions which were more general or deviated from their specific area of expertise.

Mohammad

For the most part, our students learned the training material in silos, and it is clear they didn't necessarily understand all connections nor had a clear understanding of the software development lifecycle (SDLC). They also did not start this project with a working practical knowledge of the tools needed such as git as well as the agile process. They did however have a strong foundation of programming concepts, regardless of not entirely knowing how they connect or interject.

When I was developing the curriculum for the MEng program, I made sure to keep my focus on differentiating between *teaching* and *training*. The former emphasizes the fundamentals and ensures students regularly practice them and be assessed. Alternatively, training provides self-learning opportunities during hands-on learning and projects. By incorporating both approaches, the program utilizes training packages on certain technologies that will support our graduates when they move into industry positions.

The truly innovative aspects of the MEng program come in the form of the DevShop pedagogy and the leveraging of industry partners who act as mentors and provide students with rich learning experiences at their companies. "DevShop" is a term I have coined to describe a unique process through which the classroom is transformed into a creative space, with each topic taught by developing an application. This means that students can ask questions while actively developing the app along with the instructor. This engagement in pair-programming with the instructor further helps to facilitate discussions of best practices. I have found that this approach repeatedly provides many opportunities for students to develop their programming skills and understanding of software design. I believe it is a critical aspect of the MEng program that has led to our students' success in gaining relevant employment upon graduation.

Chapter 16: Dependencies and Productivity

Friday, June 5

"The best architectures, requirements, and designs emerge from self-organizing teams."

- Agile Manifesto

-

Yazan

At first glance, Jerome and I found our authentication task very overwhelming, because we haven't really done anything like it before and had a lot of questions along the way. There were also different ways of implementing it online. Being able to communicate with Arash and the mentors made me understand it a lot better. Although a lot of issues arose from the task, we become more comfortable as we worked more on it. Our story was mixed in that the profile and the authentication had to work together. I worked on the authentication part. First, I went online, read a few websites on how it should be done, watched some YouTube videos to get more familiarized, and then planned out how I would go about that task. Being new to the task, it was scary at first to try to implement it and not break anything. But you have to push yourself to do it else it'll never get done. From then, I asked Arash for advice and spent a lot of time testing since some things didn't work the way we had thought. It turned out in the end that what we wanted to do wasn't possible, but it taught me a lot of how authentication should work with all the other database models that we had.

© The Author(s), under exclusive license to Springer Nature Singapore Pte Ltd. 2023
N. Raeesinejad et al., *The Ignite Project*,
https://doi.org/10.1007/978-981-19-4804-6_16

I would say the project is not progressing as fast as we want, but this was mostly due to the training material at the beginning, where some people had issues setting up stuff on their computers or not being fully comfortable with certain concepts. Personally, for us, when we had the issue with installing JupyterHub on Windows, we had to run everything on my computer since I was the only one with a Mac. This stopped my potential of doing other things such as tasks with the frontend which I didn't mind anyways, but this process could have been better earlier on. It didn't work for us as a team, so we moved on to JupyterLab which is working a lot faster now and our stories are going by a lot quicker. No one could've predicted this happening in the beginning, but these random barriers are the norm in software development.

"Arash and I made a promise to keep growing our beards until the MVP deadline." Mohammad declares jokingly to the team once everyone is on call.

"Hopefully with today's demo that won't be too long from now." Arash grins. After building the latest master branch on his machine, he immediately points out a couple errors that appear when running the code. I notice most of these errors have been echoed over the course of this past week due to new dependencies introduced by more complex stories that the team has begun to complete. To become better informed of dependencies, they have been reminded to update their documents whenever a new library has been added to the codebase. Arash then asks if any stories are complete in RC1 and RC2 and the entire team reports to have all their stories merged and tested. While waiting for the build to finish, he invites others to ask any questions they may have.

"Do we need to add a dependency to the package.json file or does everyone need to manually install it?"

Arash explains that each of their three services – backend, frontend, and authentication with JupyterHub - have their own dependencies. The backend dependencies are handled in a file called _requirements.txt_. Every time a new package needs to be installed inside the Docker container, it must be added with its version to the file. Next time someone brings up Docker, the new dependency will be installed automatically. The frontend dependencies are managed by _package-lock.json_, which are updated in the same way for every newly installed package added to _package.json_.

"We don't have anything automated for handling the dependencies of the authentication system yet." He continues to explain. "For now, we are just manually editing the README file for all the commands needed to setup the environments needed. I think it's better to create a file for setup so that everyone will have the same dependencies after running JupyterHub."

A WORD FROM THE MENTORS

Dependencies of the Ignite System

Requirements.txt
A file containing a list of items to be installed. In other words, it is a list of all the project's dependencies. This also includes other dependencies needed by these dependencies.

npm
Node Package Manager is a package manager for the JavaScript programming language.

package.json
Yarn looks for package.json files to identify each package and configure the behavior of yarn while running inside that package.

package-lock.json
package-lock.json is automatically generated for any operation where npm is modified. Yarn is able to import its dependency tree from npm's package-lock.json natively, without external tools or clunky processes.

Yazan starts off Team 11's demo of a user profile.:

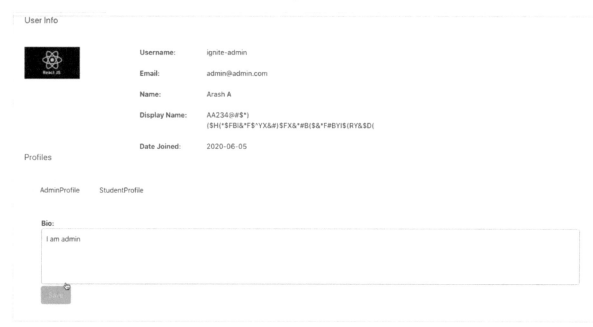

"Right now, we're only verifying the user email format by checking if there is an at sign in the email, but we could make it more sophisticated later." Yassin mentions and Arash immediately creates a new story for verifying the full user email format.

"Can the buttons on the dashboard be a little less retro?" Mohammad asks.

"It just needs some Bulma styling on top of it." Paul jokes.

Arash chuckles and creates a new story for making the global header match the style of the rest of the page. He also creates one for removing horizontal scrolling from the page.

Upon team 11 and Chelsea's suggestions, he creates another story for marking mandatory fields in the profile view, such as having a red asterisk if a field is remained empty and then showing a red boundary.

"I'm pretty sure this already exists in Bulma." Bassem notes, suggesting for the team to take advantage of Bulma's existing features.

"Are all the fields in the user profile mandatory right now?" Yassin asks.

"They are," Arash nods. "Unless Mohammad says something otherwise." Not hearing any objections from Mohammad, he moves on to the A-Team's deme of courses.

Chelsea first guides him to navigate to a course page from the student dashboard.

"We discovered some bugs that we want to correct, like having a maximum number of characters for course name and description. Also, if one of the fields are left empty, there's a post error, so we need to check those fields and alert if they're left empty." Arash creates one story for the mentioned bugs.

Chelsea then asks from clarification on the course names from Mohammad, since not all courses would necessarily be University courses and that the title could be more descriptive with more text.

"You can get some ideas from Coursera," Mohammad suggests. "For the time being, it's okay to keep the title fairly freeform and talk about the course outline in the actual description."

"We don't really have any mechanisms against what is not allowed to be in a course title and description either." Chelsea points out.

"If the platform becomes very large, we could have an id and name identifiers so that each course is unique. It's okay for now within the scope of Schulich Ignite and we can just have an admin take care of duplicate courses."

Assessing the rest of the course page, Arash suddenly points out that it is in edit mode by default. It seems Chelsea was also not expecting this behavior. She continues to point out that her team has been struggling with showing prerequisite course names instead of id's and asks for feedback regarding whether frontend column Id's are the same as primary key in the database. After Arash declines the latter, she admits they have misread their story AC, so Arash creates a couple of new stories for all their listed concerns.

"How do we add components for buttons in the user dashboard?" Peter asks.

"I prefer going to a new page instead of extending an existing page when a button is clicked."

"Can we use tabs or tiles?" Kevin suggests. Arash and Mohammad both agree that it's more modern and easier to use tiles containing icons and text within the admin dashboard. Arash creates a new story for implementing dashboard items.

"Also, everyone on the team should be on the same page regarding the format of buttons, text, etc." Chelsea says.

Arash agrees. "We should definitely finalize some guidelines."

Demoing Team Rocket's user story took less time as Paul points out that there aren't many more new things since their last demo. Nevertheless, he shows how an instructor may promote and demote a user and be redirected to new user profile page. Arash creates an additional story for creating a faculty profile when a faculty user is promoted as well as deleting the profile when they are demoted.

Thanmayee guides Arash through her and Kevin's demo of editing lessons from the faculty dashboard. They discover similar bugs regarding format and verifying empty fields as previously addressed by Chelsea along the way.

"I'm just wondering what a prerequisite for a lesson is exactly if we already have prerequisites for courses?" Thanmayee asks.

"You're right, that doesn't make a lot of sense." Arash trails off.

"What I understood is that a prerequisite could be what to prepare before a lesson." Bassem chips in.

"I think that was our original intention." Arash nods and clarifies. "In contrast to course prerequisites, where a course is a prerequisite of another course, lesson prerequisites are more like a to-do list before a particular lesson. Like reading a chapter or completing an assignment." He then creates a new story for changing lesson prerequisites in markdown format as opposed to a list of lessons.

He adds another story for the tags after noticing that they should not contain commas after being entered. After creating a lesson, he also notices a similar problem with redirecting and adds another story to be redirected to view mode after lesson creation.

"We should also implement a more reactive feeling when hovering or clicking on buttons." Arash says. After verifying for Toya that his autocomplete function for tags works, he also suggests for the dropdown to be fancier.

"We're using a lot of tables right now," Kevin voices his concern. "How will we be implementing the website to work for mobile view?"

"By having less columns in tables." Arash answers.

"Would admins and professors really be using the system in their phones?" Chelsea asks speculatively.

"It is a possibility."

Chelsea starts to ask another question, but Yazan beats her to it, asking whether errors may be caught on models instead of hardcoding the frontend. Arash explains that while this is possible, the latter option is better. Errors should be prevented in edit mode, but the team should also check for them in the backend. After getting an "okay" from Yazan, Arash asks Chelsea if she had a question.

"Steph cleared it up for me in the chat. I must have missed something because we keep adding so much stuff."

"Do we have any templates or guidelines for the containers and style or formatting?" Stephanie asks.

"I will talk to Masoud about creating one."

"We can make one." Masoud agrees. "In the meantime, the team can just copy and paste the formatting they have right now."

"The template can include other components as well, such as alignment search bars and lists." Arash suggests. "Generally, it's good to have guidelines for common features."

Lastly, Bassem demos how to maintain a course offering on his own machine:

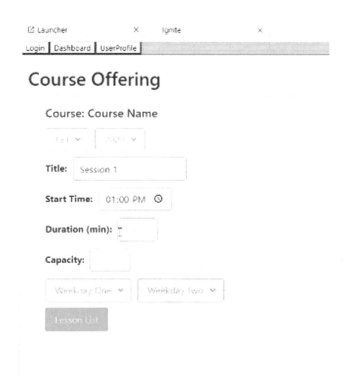

Arash repeats the same comments regarding formatting as for previous stories and Bassem assures that he has been taking notes and will implement those standardizations. He follows with some clarification questions regarding the number of lessons per week and the dropdown of pre-selected years, and Mohammad clarifies his requirements.

"Should we have a set color scheme for the whole website?" Thanmayee asks.

Bassem is quick to voice his concern. "I don't think we should worry about making things pretty before the guidelines are set."

"The text is default for now, but we should set a color scheme." Arash agrees and creates a new story for unifying colors for elements such as the header, buttons, dashboard icons, and alerts. The entire team spends the next couple of minutes establishing which colors to use for alters together, with consideration of existing Bulma tags. They finally finish all demos and grooming stories.

Arash states that for the next sprint, the team can either focus on stories for all the bugs found or implement new features, keeping in mind that they are technically behind on schedule by two weeks. To deal with their time constraint, he suggests reducing the priority for some of the minor bugs and lists some features that the MVP could be built without. Mohammad agrees that they will implement "nice-to-have" features after the MVP deadline. After a couple of minutes of Arash and Mohammad determining the priority for each bug, the team estimate their sizes in the chat, which are mostly comprised of 1's and 3's.

With this meeting having lasted for three hours, Arash promises they will have a new grooming session every other week so that demos don't last so long in the future.

Thanmayee

The main challenge with completing our task this week was that different people designed the backend and the front end of the code. It seems like the developers who developed the backend were not aware of all the fields that the front end needed. It was challenging to implement the major changes needed to be made as Kevin and I did not develop the backend and not that familiar with its structure. There were lots of merge conflicts with the tags section, because Toya and others from the team were working on the autocomplete of tags, while we were looking at the lesson and tag model as a whole. Therefore, it took a lot of effort to merge the two codes to ensure that everything works. We were not familiar with any frontend design, including important details of functional programming such as how to optimally structure your code, using React hooks etc. For this reason, we found that we needed to refactor the code multiple times in the process.

As we kept going through the task, it seemed to be getting bigger and bigger unexpectedly. For example, we spent at least 2 days fixing the backend code. Masoud's lesson page from the tutorial will need to be fixed so that all the fields we added can be displayed (we were not able to complete this yet – it will need to be another story). There were still technical issues with getting JupyterHub to run as expected and this took up at least 2 days in the past two weeks as well. It was really challenging to be productive because of the following these tasks are generally new to us and on top of learning and coding in new frameworks, we were also dealing with many technical issues with simply getting things to run and connecting the frontend and the backend. We found that we spent a lot of time fixing stories that were supposedly complete because they were not working as expected.

It also feels like with so many of us working so closely on this project, we are always either stepping on each other's toes or waiting on other people to finish their tasks before being able to complete our own. I hope that as time goes on, we can get into more specific and contained portions of the project with our team, so that we can be more productive. I expect productivity to increase next week, because there was a lot of learning to be done with this story since it was our first real front-end task. Now we are much more familiar with what we can and cannot do in React, and we expect other front-end tasks to be similar. As well, hopefully we are over the hump with technical issues now so we can focus more on actually programming.

Chapter 17: Refactoring

Tuesday, June 9

"The word refactoring should never appear in a schedule. Refactoring is not a story or a backlog item. Refactoring is not a scheduled task. Refactoring is immediate and continuous. It's like washing your hands in the bathroom. You always do it."

- *Robert C. Martin*

Today is the extra grooming meeting that Arash scheduled for the team in lieu of stretching out future demo meetings to accommodate for grooming upcoming sprints. Since last week, more members admitted having purchased MacBooks to be able to set up JupyterHub and JupyterLab. Everyone's focus this week is mostly centered on fixing urgent bugs that were discovered during the last demo meeting. To speed up the PR reviewing process, Arash has directed the team to submit smaller chunks of work instead of waiting a long time for a review, since smaller PRs have faster turnaround periods.

Arash kicks off the grooming meeting with cleaning up tasks in Team 11's project board, asking whether each one is tested.

"When PRs are merged, their issues are closed automatically. Please close merged issues yourselves in the future to save time on my part."

While their biggest story, ⓘ`User Profile`, revealed to have some bugs, it is moved to `Accepted` with the smaller stories for meeting all functionality requirements. Any finished stories waiting to be demoed at the end of this sprint is moved to `Done`. Looking at the reduced number of stories left to be implemented this sprint, Arash concludes that Team 11 needs additional tasks for the remainder of the week.

He proceeds to repeat the same procedure for Team Rocket and A-Team's respective project boards. After confirming which of the teams' tasks have been merged and tested, he moves them to `Done`. At least half of the members from both teams will need something to for the remainder of this sprint as well.

"I'm still a bit iffy on the relationship between RCs and MVP." Mihai admits to Arash once he finishes cleaning up the boards.

Navigating to *Milestones,* Arash starts to explain that release candidates are small and achievable so that results may be viewed sooner.

"We initially planned for RC1, RC2, and RC3 to become the MVP. RC4 was outside the scope of the MVP, however after some discussion about what different teams would like to work on, a part of the *Course Management* and *Couse Offering* stories from RC4 has now been added to the MVP." He pauses for the team to take in their visualized progress for each RC:

"Ideally, a release candidate is generated after every sprint, which as I said before must be deployable and working code. Because of some hiccups in the beginning involving everyone's learning curves and technical issues with setup, the RCs are not matching with sprints one-to-one."

He further notes that RC1 is done for the most part and they have already briefly discussed the requirements for RC2 last week. It is now Arash's plan to assign one or two teams to RC3 and one team to RC2. On that note, they proceed to go over all the stories in the backlog to be implemented this sprint. Similar to recent grooming sessions, for each story that he presents, Arash invites the teams to ask any clarification questions and voice their own opinions regarding the requirements and functionalities.

After a couple rounds of discussing and sizing stories, Arash teasingly points out that there seems to be some friendly rivalry between Team Rocket and A-Team for getting assigned to more frontend-oriented stories. For the last story, their size estimates are neck-and-neck between "5" and "8" in the chat. Since most of their reasoning for the latter pertains to unknown tasks associated with the story, Arash instantly decides breaks it down into smaller stories.

"While eight as a number does not have any meaning, it is large in the context of this team, so I don't want any stories to have this for their size. My strategy is to make stories easy enough to be ready for the MVP at the cost of later refactoring."

At the end, he tallies up the total size estimates for each team this sprint. For example, A-Team will be working on a 16-point sprint this week. Along with Team Rocket, they have more than enough work. Arash expresses the challenge of assigning tasks to Team 11 in contrast while attempting to simultaneously avoid team dependencies. Instead, he assigns them to RC3. Following their previous spike regarding adding video streaming for live lessons, they will now try to implement one of the potential solutions they have found.

Accordingly, he creates a new issue for RC3 named ⓘ**Add Video Streaming**. The team spend some time team discussing the pros and cons of video streaming through YouTube versus Zoom. Since Team 11 needs something to work with, Mohammad decides for them to work with Zoom for the time being. The plan is for them to first attempt adding a Zoom SDK sample application to the new JupyterLab extension and conclude which platform to use after two days.

"Please remember to only work on tasks that you are confident in getting finished by the end of the sprint." Arash reminds. "It's better to have something deployable by Friday instead of several unfinished tasks."

Wednesday, June 10

For their standup meeting, Team 11 reports to have started breaking down their assigned RC3 stories into tasks with a promise from Arash to find more tasks for them if there aren't enough for their whole team to work in parallel.

"Does the product owner want to have all the features of Zoom SDK such as the high resolution?" Yazan asks.

"You can give it a try and see how good the resolution looks." Arash answers and thanks Yazan for bringing it up.

On the topic, Mihai, on call as usual, starts asking a series of further clarifying questions regarding live streaming lectures. Firstly, he confirms with Arash that the video may include both the professor's face as well as screen sharing and that students may type along on the platform at the same time.

"In that case, the screen would need to be zoomed in so that students can see the professor's face on the screen." Mihai points out.

"I understand your point." Arash says as he pulls up their system prototype on his screen. He opens several tabs for notebook, video, chat and shows different ways in which students can split their screen into tabs according to what they want to focus on.

"I didn't know that users could split tabs until after you showed it." Mihai reveals and they conclude that splitting and extending tabs negates the issue of having to zoom the video.

With no more questions or comments from the team members on call, Mohmmad double checks with Team 11 if they are confident enough to finish their stories in time and they confirm their confidence once again. "Despite your progress so far, we need to lift this off the ground since things are a little more time sensitive now."

Thursday, June 11

True to their word, Team 11 has started working with samples on GitHub and will try to implement video streaming on JupyterLab today.

In Team Rocket, Mihai, Thomas, and Paul are planning on working on the advanced search feature. Meanwhile, Kevin and Thanmayee are working on additional requirements for the lesson page. They had some issues with React hooks but will talk with Masoud today during his office hours and are expecting to finish their story today.

Peter is the last to report. "I worked on the dashboard bug fixes and submitted a PR and will continue to fix more frontend bugs like the icons. There's a lot more refactoring for the course list story, so I had planned to have that in another PR."

"Sure."

"But," he continues. "I had Master merged in my working branch yesterday, and it seems like someone already fixed the code for me, which I'm thankful for but I don't know who it was."

"Fortunately, GitHub can figure that out for us." Arash grins, navigating to the specific file that was changed.

"It sounds a bit negative, but the *Blame* feature provides a history of which lines were changed as well as when they were changed by whom in which PR, whether this change was good or bad."

In A-Team, Michael and Bassem are working on adding recurring days to the course offering page. Stephanie reports next that she will work on the ability for an admin to upload their icons and resize their avatars. Chelsea will also work on the backend portion of uploading the emails to the database and Toya will develop the frontend for the user profile, for the user to select existing avatars or get assigned a default icon.

Arash directs all their questions for clarification regarding restrictions and dimensions toward after the standup portion of their meeting. He briefly mentions to have configured a Django CI to run automatic tests for the backend to resolve the backend issues they were facing yesterday.

"If you are working on a particular part of the backend, please make sure you fix all the link and test errors before submitting your pull request. Moving forward after this sprint, it won't matter whose code breaks the pipeline; if your PR is rejected, you are responsible for fixing it. At the end of the day, the whole team owns the code, and we are all responsible for this code base."

A WORD FROM THE MENTORS

Within an agile team, **shared responsibility** pertains to making decisions together with acknowledgement of carrying out the consequences of such decisions. This work dynamic inherently rules out any notion of individual tasks, rather encouraging team members to refer to them as "our tasks." As the entire team becomes more involved and feels the weight and impact of shared responsibility, the amount of work hand-offs will decrease. As a result, the team's progress will not suffer from unnecessary waiting and knowledge transfers from one member to another.

The philosophy of shared responsibility is mainly oriented towards fast value delivery by the entire agile team; this consists of everyone who helps create value, including the product owner and anyone involved with development, testing, operations, etc.

Friday, June 12

After exchanging their usual greetings in the beginning of their demo meeting, Chelsea points out that there is something wrong with running the Master backend, but the other members say that they're able to run it just fine, suggesting this may be occurring because of her machine's environment. Arash starts pulling from the Master and builds the backend on his machine, having seen some 404 errors himself earlier. The next couple of minutes are filled with silence and the occasional typing sounds from Arash's end.

Last to join the call, Mohammad grins and instantly asks, "Does the team have a lot of demos to show today?"

After a short period of awkward silence, Arash silently says they will see, still typing.

It seems the team experienced a lot of problems with linking this past sprint, so Arash stresses the importance for them to understand how to configurate linking in the backend and ensure that their PRs are not causing merge and migration conflicts. He explains that the general practise is to have small PRs in order to not be behind on the most updated branch by too long. Before submitting every PR, the team member must test and then merge their code by either pulling the Master branch and then pushing or remaining in the branch, searching for Master, and then merging.

"Migrations are not supposed to be changed." Arash explains. "However, if they need to be changed, a new migration file is created, which I only want to see in a PR. Think of it this way: We can't just ask our customers to delete their entire database and start from scratch every time a migration needs to be changed or added to our system. To bring a customer to the final state, all they need to do is incrementally add migrations."

A WORD FROM THE MENTORS

Version control systems are all about managing contributions between multiple distributed authors (usually developers). In cases where multiple developers are editing the same content – such as changing the same lines in a file, deleting a file that was in the process of being modified by another developer – a **merge conflict** may occur. In these cases, Git cannot automatically determine which modification is correct. Instead, it will mark the file as being conflicted and halt the merging process. It is then the developers' responsibility to resolve this conflict.

Migrations are Django's way of propagating changes made to models in your database schema (e.g., adding a field, deleting a model, etc.). Similar to a merge conflict, a **migration conflict** occurs as a result of multiple developers altering the same Django model in a project.

Next, Arash prompts everyone for their updates on their stories to see if there are any that can be demoed today.

Team 11 has got the Zoom widget working for the most part, and just need to make sure their components are in the right place so that adding features next time won't be difficult. They need to have a talk about this with Masoud next week. Since they haven't created a PR yet, they cannot demo today. Similarly, the members from Team Rocket have finished and tested nearly all of the stories for fixing bugs, however their new story, ⓘ**Story Lesson Prerequisites** has not been tested yet, so Arash tells them to hold off on its demo.

"We did test each other's code within the team," Peter reveals. "But I'm not sure if that would be sufficient."

"As long as someone who hasn't developed the code has tested it, then it can be demoed." Arash explains as he moves the story to `Done` and cleans up the rest of their project board.

Lastly, A-Team has fixed all their bugs for ⊙**Maintain Course Offering** which has been tested by another group and ready to be demoed. The changes from last week were not implemented, because the team assumed that they belonged to a separate story.

"Also, our story for the course name restrictions was working last night, and it was merged and tested, but then it caused a lot of backend issues in the Master." Chelsea reveals and Toya echoes her report.

"Usually, as long as someone tested it and it worked, it is considered to be done in the developer's point of view and it doesn't matter what happens afterwards." Arash explains assuredly.

"What exactly needs to be in the done column?" Peter asks.

"Tasks are only for the development team and to keep track of who's doing what, but what the product owner cares about and needs to be tested are the stories with the full features. So, for example, in my point of view, the done column should contain only stories and bugs to show new features as well as existing features which have been fixed."

It takes some time to install a new environment and build all the files in order to commence their demos. In the meantime, Arash tells the team that their code implemented on JupyterLab should also run on JupyterHub. He then navigates to all their open issues in the backlog, explaining that refactoring is sometimes easy and other times not. Usually when it comes to planning sessions with the stakeholders involved, the latter group always wants to see the full feature working, so the development team ends up spending more time developing features rather than refactoring previous code and polishing it up. As a result, refactoring keeps getting pushed back in the backlog. He tells the team that it is much better, if a member is already in the code, to refactor it right then and there, even though it might take a while.

A WORD FROM THE MENTORS

Refactoring is the process of restructuring software internally to make it easier to understand and cheaper to modify without changing its observable behavior. Rather than cleaning up code, rewriting code, or fixing bugs, refactoring is a controlled technique for improving the design of an existing code base. Its essence is applying a series of small behavior-preserving transformations, each of which are "too small to be worth doing". However, the cumulative effect of each of these transformations is quite significant; it reduces the risk of introducing errors and avoids system breakage in the process.

Refactoring is often taught in the context of **Test-Driven Development (TDD)** that is often described in terms of the **red-green-refactor** cycle:

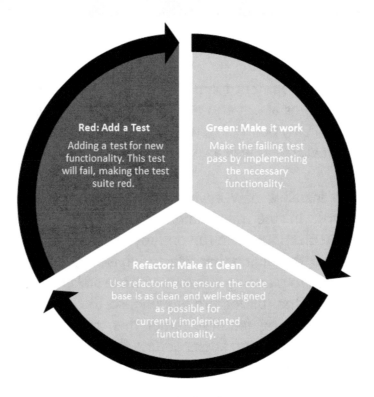

Benefits

Maintainability	It is easier to fix bugs because the source code is easy to read and understand.
Extensibility	It is easier to extend the capabilities of the application if it uses recognizable design patterns.

When to refactor

In computer programming, a **code smell** is any characteristic in the source code of a program that possibly indicates a deeper problem. Code smells are usually not bugs; they are not technically incorrect and do not prevent the program from functioning. Instead, they indicate weaknesses in design that may slow down development or increase the risk of bugs or failures in the future. Bad code smells can be an indicator of factors that contribute to technical debt.

Common code smells include but are not limited to the following:

- Duplicated code
- Shotgun surgery
- Large class
- Inappropriate intimacy
- Feature envy
- Long method
- Too many parameters

Review www.refactoring.com/catalog for more information about Refactoring Catalog.

Since the backend is still building, he moves on to discuss tasks created by the teams from their assigned stories. He converts some tasks into stories after deeming their sizes too large. Since Team Rocket and A-Team have a lot on their plate, Arash assigns a new story to Team 11. after Arash gives a rundown of its general workflow and acceptance criteria, they discuss the user experience aspects of a story involving an instructor enrolling students and TAs in courses which requires Mohammad's attention

"I'm having a hard time visualizing it," Mohammad confesses. "When you talk about software development, it can be relatively abstract, so although it sounds like it makes sense now, it may be missing something once shown either through a diagram or demo."

With less pre-established and enforced acceptance criteria, the rest of the team proceed to suggest further features. Depending on the approval of Mohammad and Arash, some get rejected and others become prioritized as later features down the road.

Once the backend finally finishes building on Arash's machine, they switch gears back to the demo portion of their meeting. After pulling up their system on his screen, he notices there is an inconsistency in styles across components belonging to different stories.

"I created a new story for unifying the styles like you wanted, Arash." Peter says. "Though I think it would require some rework on Team Rocket's part."

"I think we should follow the proper React way for our own experience as frontend developers, regardless of the implied rework." Paul advocates.

"I won't try to block any of the team's ideas." Mohammad assures, saying that things look good for the most part. Regarding the colour display, they may seek input from someone experienced in graphic design. He plans to have a long effective talk after the MVP deadline, either with Masoud or someone else who is experienced to set a theme for the team to follow going forward.

After their short demo, Arash opts to not create any new stories for bugs this time, trusting that the teams are already aware of and will fix them while working on recurring stories this coming sprint.

"Should we create a Slack channel for bugs that we found?" Bassem suggests.

"You can just label issues as bugs on GitHub and have discussions under that issue instead." Arash says. "This way, you have all the background information for the bug in one place which helps with sizing similar issues in the future."

"Do we have any plans for deployment?" Mohammad asks Arash.

"I have planned to deploy this on Google Cloud, because it offers service for JupyterHub, where we would just need to create an instance and add our own extensions." He then lists some milestones which can be closed by deployment, estimating that by the end of the next sprint, 90% of the MVP will be complete.

"Hopefully, it will be better than this."

Toya

Peter has asked for help on refactoring the Course frontend on the admin dashboard. I decided to help as much as I can since I played a major role in developing the frontend for it. There has been a lot of merge migration conflicts appearing in the backend now. Linting appears to not be the user and is causing additional test checks to fail. This means more linting checks now need to be done for backend developers before PRs are approved. It seems that our changes are being overridden by resolving merge conflicts. This is a problem since now things are breaking during demo that no one was prepared for. There are too many bugs being found and passed to me and my team which is preventing us from working on other stories. Also, it appears that bugs show up in our stories that weren't caused by us. It's frustrating because it makes it seem like we haven't done anything when we've been actually fixing bugs caused by others and neglecting our own stories.

Chapter 18: Overcommitment

Monday, June 15

"Productivity is never an accident. It is always the result of a commitment to excellence, intelligent planning, and focused effort."

- *Paul J. Meyer*

While waiting for Arash to join their standup meeting, Mohammad asks about Team 11's progress with video streaming, and they report that this feature hasn't been implemented yet due to some difficulties experienced with CSS styling. He follows up by asking how they like working with Zoom so far and Yazan reveals that it involves several unknowns regarding its implementation with JupyterHub.

After a couple seconds of silence, Arash finally joins the meeting. "Sorry I'm late, have we already finished the standup?"

"We've just been chatting."

Yazan volunteers immediately to report that Team 11 have been trying to finish integrating Zoom and will need to discuss with Masoud how to add a React component for entering meeting codes.

"Whenever we run Zoom, it takes up the whole browser window and we've read on forums that it doesn't support resizing." It seems they will need to figure out how to override CSS to resize it on their website. Hopefully, Masoud can help them with this issue as well.

"Have you guys integrated Zoom in the extension or just doing tutorials for now?" Mihai asks.

"Right now, we have a separate widget with all the features and we're still trying to make it work." Yazan answers.

"Are you interested in helping out or just asking in general?" Arash asks Mihai, who replies while it is the latter, he still wants to do the tutorial in parallel with Team 11. Arash says that he can have access to their code if he wants to help with this task.

"Once Team Rocket starts burning through our tasks, some of us will probably take a look at this one with Team 11." Mihai notes and Arash agrees with their course of action, reminding Mihai that their own team tasks are still their top priority.

Wednesday, June 17

This week marks the end of the Spring term as well as the MVP deadline, meaning that the MEng student interns have reached the half-way point of their projects in Mohammad's class. Thus, A-Team and Team Rocket will have their break during the following week before the start of the Summer term. Each team was set to finish most of their stories prior to their extra grooming meeting which took place yesterday afternoon to address two-weeks' worth of work to be assigned. During that meeting, I noticed that Team Rocket members were generally first to call dibs on the majority of stories to which Arash asked who wanted to be assigned after breaking them down. I didn't expect this level of drive from them towards taking on more work since they had also previously shown a keen interest in collaborating with Team 11 to implement the live video streaming feature. Their usual reasoning behind intending to take on a new story stemmed from their previous experiences and existing knowledge and skills, as I heard members frequently justify that they should work on a particular story because they have done a similar story before and would thus be able to complete it faster.

The teams seem to be adopting a relatively new dynamic in contrast to when everyone was working on many new things for which they did not have much experience during training week and earlier sprints. Due to the large number of developers and increasing number of more complex stories which needed to be groomed in a limited period of time, it seemed, with the exception of Team Rocket's overwhelming claim on frontend-related stories, that most members were not given many chances to truly contemplate on whether or not they wanted to be assigned to a new story. Stories were also assigned on a first-come first-serve basis in lieu of using tactics like story points or voting, which were introduced by Arash in the very first kickoff meeting and never practiced to my surprise.

Arash's voice suddenly pulls me from my thoughts and I focus back on the current standup meeting. "Do you guys feel confident about finishing your stories?"

"I'm not very confident in finishing all the stories for Team Rocket." Mihai admits. "We were not expecting to run into problems. We were very close to finishing our initial story for the advanced search feature yesterday, however we were derailed by a very small piece of code."

"One of the skills you need to know as a software developer is knowing how to not overcommit." Arash smirks.

A WORD FROM THE MENTORS

Since development teams should ideally meet all their commitments, there is a cause for concern if they are regularly over-committing. **Over-commitment** means committing to more work than what can be realistically done. This inevitably leads to lower productivity and quality, team burnout, and failure to deliver as promised. Consistent over-commitment is typically a result of one of the 2 following reasons:

- The team is unable to properly assess their velocity.
- The team is pressured by external forces to over-commit.

To mitigate over-commitment, you must first be aware of its key symptoms, which include but are not limited to

- Overwhelming work in progress (**WIP**) - Work that has entered the development process but is not yet finished
- Too much task switching
- Many "not Done" backlog items at the end of sprints

Listed below are common solutions for mitigating over-commitment, based on principles from the Agile and Extreme Programming (XP) communities.

Solution	Definition
Yesterday's weather	An XP principle which states that "you'll get as much done today as you got done yesterday". In iterative projects, you must plan to complete the same amount of work as that of the last iteration.
WIP Limit	WIP should be limited by the maximum number of in progress work items within a system; this is based on **Little's Law**, $$WIP = Throughput \times Cycle\ Time$$ where **throughput** is the system output measured by average arrival/departure rate of the work item and **cycle time** (or lead time) is the average time an item spends in the system. In Kanban, WIP limits are applied to each process state, mandating that outstanding work must be completed before new tasks can enter a process state.

Swarming	A behavior whereby team members with available capacity and appropriate skills collectively work (swarm) on an item to finish what has already been started before moving ahead to begin work on new items.
Spike	An XP term referring to work whose primary purpose is to explore potential solutions or otherwise gather information. This is often an effective way to proceed forward when encountering uncertainty.
Small user stories	Good stories tend to be small, with sizes typically equal to a few person-days. A good rule of thumb is that any single story does not take more than 50% of a sprint; for example, for a two-weeks sprint, a story won't take more than 5 weekdays. Anything beyond this range should be considered too large to be estimated.

Peter

Today I have been working on my first task contributing to the Django backend REST API. From the sprint grooming session on Tuesday, we initially rated this backend task a score of 3 (easy). Looking back, I did not fully understand the scope of the story and therefore rated the story much less difficult than I did during the grooming session. I worked with Thomas and Michael for many hours today trying to figure out how to insert a new model object into only a certain type of User account. We have only been able to insert the new Enrollment model into all User accounts which is the parent class for Students, Faculty, and Administrator users. What I am learning right now is that my skills in frontend development are much stronger than that in the backend.

Friday, June 19

"My laptop is so slow." Arash complains after waiting for several minutes to build and load their system for today's demo.

"You can join our new laptop club." Peter jokes.

Chuckling, he opens Team 11's project board first. It seems they only have one task `In-progress` with everything else in `Done`. There are also a couple of bugs left to be fixed for the video streaming feature.

"Would Zoom work with JupyterHub?" Yassin asks after his team confirms with Arash that they can have all their stories demoed for today's meeting.

"We'll see after we do everyone else's demos first." Arash assures.

Team Rocket, to no one's surprise, still have a couple tasks `In-progress` and `To-do`, as a result of overcommitting to multiple new frontend stories, becoming involved in Team 11's video streaming story, and spending substantial time on fixing unexpected bugs or refactoring their frontend tasks to reflect some backend changes in their originally assigned stories.

Arash proceeds to demo Team Rocket's stories, first checking how to edit date fields for creating lessons.

"Should I be able to create a lesson in the past?" Arash asks inquisitively.

"We haven't included any error checking for that so far."

"A course offering has a start and end date. To create a lesson, there should be error checking that the lesson date is between those dates and must happen on either a Monday or Wednesday."

"What about the case of scheduling workshops or any other events occurring outside the scheduled lesson dates?" Kevin brings up.

"We could have something like a session." Peter suggests.

"This could be a UI definition, which the day of the week restriction won't apply to, but instead it can call something different from the backend." Arash ponders, then leaves it as something they could implement in the future. For now, they'll definitely need to enforce is having a lesson date be within the start and end date of the course offering.

"It would be useful to do that." Mohammad agrees.

Another bug that comes up is the requirement for each lesson to have at least one tag, for which Arash creates another issue.

"Some things were not rendering properly on the website even though we worked on Mihai's computer and my own." Paul mentions. "We will have to look into it to see if it was due to a migration issue."

"The website looks good despite not balancing the styles." Mohammad comments.

"We worked really hard on making sure that the website looked nice but it's not showing everything in the demo." Kevin says regrettably. This may be a formatting glitch that Team Rocket will need to resolve.

"Let me load the site on Firefox instead." The features are shown a lot better this time, with a more unified color scheme and alignment of widgets. After Arash and Mohammad approve of the look, Mihai suggests that the members could also demo on their own machines. In the meantime, however, they continue demoing on Arash's screen.

Next, Mohammad suggests changing their wording from "season" to "semester" when creating a new lesson page, once again inspired by the University offering templates for each series of days for instructors to schedule their courses. He asks if it would be possible to implement this dynamically so that an instructor can select specific days for a course offering.

Bassem jumps in to request for everyone to wait until his team is done implementing checkboxes for all days, since they only have the basic functionalities done for now.

"What Mohammad is saying is to keep the same backend but just show it differently in the frontend." Arash clarifies.

"Then that's easily doable." Bassem says in a relieved tone and Arash creates an issue for their course offering terminology changes.

Than and Kevin are next to demo their enrollment page:

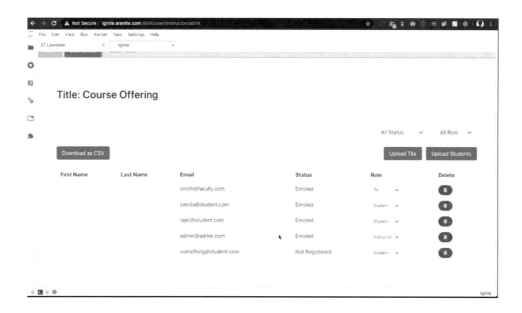

After once again expressing his approval for the look of the page, Arash creates some bug issues involving not creating new users in the enrollment page and correctly implementing their search by email feature.

He moves on to A Team's demo next involving the creation of courses. He logins with the admin credentials and goes to the Course page.

"We may have created a bug in the last sprint which was not prioritized for this sprint." Toya points out and after identifying the bug, Arash creates a new issue for them to match the Course page form with their guidelines on enforcing mandatory fields. For example, they should implement red asterisks for fields that should not be left blank by the instructor when creating courses. Lastly, Toya mentions he had helped Peter last week with refactoring page handlers for consistency, so he has transferred its story to their own team's project board.

With no additional remarks from the rest of the A-Team members, Arash moves on to ask Team 11 if they can demo anything in the current setup on his machine.

"We can demo the mandatory fields for editing profile and affiliation in the profile page." Yassin suggests.

Arash navigates to the user profile page and frowns. "The fields are not aligned properly."

"I couldn't get CSS to work properly even though the functionality is right." Yassin admits.

"Did you ask for help from Masoud?"

"No, but I should have." he says regrettably. Arash creates a new bug for him to fix and Kevin offers to help him with it.

Team 11 has three more stories in Done which cannot be demoed in the current setup. Arash stops his JupyterHub and begins to run JupyterLab on localhost instead. Yazan explains he has scheduled a series of recurring Zoom meetings ahead of time for the purpose of this demo. After Arash enters the meeting ID and password and clicks on meeting attendee, a Zoom video call appears on the screen.

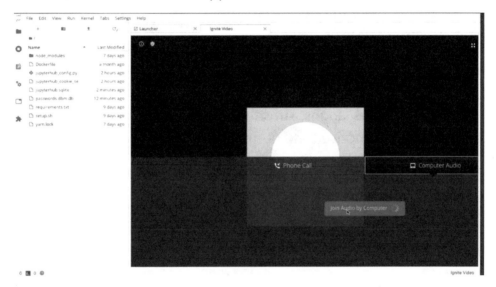

The video streaming tab is nicely structured in one tab, while the other tab is left empty for the purpose of students to be able to take notes and code along in the middle of a live lesson.

Stephanie:	**This is awesome! Great job boys.**
Mihai:	**Agreed! Good job guys!**

"It does look great, but I don't understand why a user would choose if they're a meeting attending or a host. They should just go into a meeting." Mohammad points out.

"Definitely, we will look more into it." Yazan agrees. "We are also currently looking into authentication for Zoom's chat feature."

With the demos done, Arash quickly leaves the call to steep his tea. In the meantime, Mohammad tries to engage in some small talk with the team. He asks jokingly if he looks skinnier at all, and everyone laughs. Arash comes back just in time before the silence becomes too awkward (I've noticed the normal threshold is around five seconds).

With the end of this week's demos, Arash and Mohammad seem very impressed with the teams' progress.

"Thank you everyone for your hard work, we are very close to the MVP! Looks like I'll be shaving my beard soon." Mohammad laughs.

"Agreed." Arash nods and wraps up the meeting by announcing that they will have a grooming meeting on the Monday after the MEng teams' week-long break. For Team 11, who will not have a break as full-time undergraduate interns, they will have their own standup meetings next week as usual. Lastly, he shares the general features that are coming up which have been assigned to teams based on their interests.

After MVP:

- Frontend theme and UX design

A-Team
- Document sharing between TA and student
- Course management and gamification

Team Rocket
- Start a lesson / view the recorded lesson
- Analytics dashboard

Team 11
- Chat in Ignite

Mihai

This week, our team split into pairs and collaborated between each other when necessary. I worked with Paul on getting the advanced search component to meet the acceptance criteria and implement it in the Lesson List and Course Offering List. Paul and I used VSCode Liveshare and Discord screensharing to pair-program, with one person coding and the other supporting by checking code, searching the internet for answers, and providing ideas and suggestions. Every few hours we switched roles. This was a very effective way of approaching the tasks and I think we would not have met the deadline otherwise. What I learned is that importing a library can make development easier and the code cleaner but doesn't give you the same flexibility as writing the component from scratch.

One of the challenges was getting a component to do what you want it to do when it doesn't have the attribute to control it; We got stuck on a very, very small piece of the code - the autocomplete functionality was there but due to the nature of hooks refreshing the component on state change, the autocomplete results were not showing up. Paul and I tried a number of approaches and consulted Masoud who also tried some things but none of us were able to fix it. Masoud mentioned he may try to look into it over the break but suggested we push the code since the functionality is 99% complete. Paul and I then created another hook for course offering based on the advanced search hook for lesson list. Both of the stories were completed in time for the demo on Friday!

This coming week is a break between the spring and summer semesters. My hope is that I can finish the React course on Udemy. Once I finish that I may start the Typescript course that I also have on Udemy.

Chapter 19: Effective Communication

Friday, June 26

"Effective communication helps to keep the team working on the right projects with the right attitude."

– Alex Langer

Throughout this past week's sprint without the MEng teams, Team 11 primarily worked on fixing functionality bugs and building a chat UI from scratch for the Ignite platform. I expect today's demo meeting to be a short one, but you can never know, given the number of unknown elements that Team 11 reported to have struggled with during the last couple of standup meetings.

After quickly demoing a rather simple-looking UI for the chat feature, Team 11, Arash, and Mohammad begin discussing the prospect of a global chat during a live lecture, with Mohammad earnestly expressing its usefulness. "When you send a group to a TA, then they should only talk to the TA and not the instructor which causes confusion."

"In that case, we would need to implement the functionality of assigning students to TA's." Arash notes.

"Zoom does this with breakout rooms, where for example every five students could be assigned to a breakout room, but then as an instructor, I would have to keep jumping to every room to check. It's good to teach the students for a period and then have them chat with the TA in their breakout rooms." Mohammad takes a pause to take a sip from his teacup. "Would it be possible to use Zoom for the global chat and have our own chat as well?"

"That would be too complicated," Arash grimaces. "We should just choose one."

Mohammad leans back on his chair and crosses his arms. "A global chat would be useful when an instructor is addressing all the students. It should be possible to disable global chat and have students talk to their TA's instead. With what we have right now, I believe that it's better to implement our own global and breakout chats." He seems very attached to the breakout room features, even though there's a delay for the instructor to jump from room to room.

Mohammad then proceeds to demonstrate how breakout rooms work on Zoom. As the host of this meeting, he starts assigning everyone in the meeting except for me to a breakout room. All the participants start disappearing one by one and I'm left staring at a blank screen for a couple of seconds before they all return to the main meeting.

After once again stressing the importance of interactions between TA's and students, Mohammad manages to convince Arash to finally agree on adding the feature of having a webcam for the TA's later, given that this is possible with the Zoom SDK. With that in mind, he creates new stories for Team 11 to implement the features they have discussed thus far, including a global chat, randomly assigning TA's and students to breakout rooms, and video chat changes based on the distribution of rooms.

"Is there is a way to communicate between two extensions, since the chat and Zoom are on different ones?" Yassin asks.

Arash shakes his head. "It's impossible, but you could use the backend to send information between extensions."

"If a TA becomes a co-host of the breakout room, I think they should be able to screen share." Yousef notes. Humming in agreement, Arash quickly makes some edits to their stories' AC.

"I'm just trying to picture what everybody can see." Mohammad squints. "If you go into a breakout room as a student, you see your TA on one side and your code on the other. Would the TA also see the students on one side if they choose to have their webcam on, and on the other have their own screen for document sharing?"

"Document sharing is not a feature on Zoom." Arash deadpans.

"I'm referring to the tabs on our system, not just Zoom. Would the TA see a bunch of videos of students' codes that they can jump to?"

Arash closes his eyes to ponder for a couple of seconds before answering, "If the student decides to share their notebook, the TA can then have access by editing and saving that notebook. Aside from this, the UI would look the same for both TA's and students."

Mohammad nods and tells them to continue discussing while he draws a diagram of what he's picturing on his iPad. In the meantime, Arash and Team 11 discover a couple more bugs in their chat UI, resulting in Arash declaring that he won't mark the chat features as done for the time being. Before they can add more stories in addition to those bugs for the next sprint, Mohammad quickly shares his screen to show his diagram for what a TA would see on their screen:

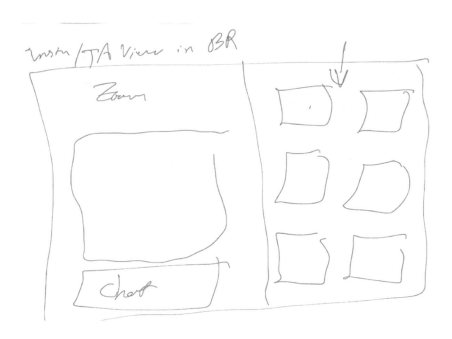

"The TA can see what each student is doing as an overhead, and if a student asks a question and grants access, the TA can edit their code." he explains. "I want the instructor to see this same UI in a breakout room."

"Sounds good, but this would be a feature for after September." Arash infers and Mohammad reluctantly agrees before they wrap up the meeting.

Tuesday, June 30

Today marks the last day of June as well as the return of the MEng students from their week-long break. Apparently, Arash is not available for the entirety of this week, so the teams have been encouraged to seek help from Yousef and Masoud instead if needed.

Since Mohmmad is not present for today's standup either, Yousef asks each team to report what they have done and will work on today without displaying their project boards.

"We've been working on the UI bugs from last week's demo and have pretty much finished them up." Yazan reports first on behalf of Team 11. "We're now starting to work on the chat feature UI."

"Thomas and I are still working on our tasks." Peter says. Likewise, Kevin and Than report that they will soon wrap up on fixing their bugs.

"Paul and I are working on adding an advanced search." Mihai says.

"We'll submit a PR soon," Paul adds.

In the A-Team, Bassem and Michael have completed their tasks, which are now merged, and they need someone to test their frontend.

"Since Mohammad added some more functions to our story, we will create more tasks for them this week." Bassem includes.

In contrast, Stephanie, Chelsea, and Toya do not have much work left for this sprint, aside from testing stories assigned to other teams.

"Okay, thank you everyone. Please notify me directly if you don't have any tasks. I will host another meeting later today as a mini grooming session for those who need more work."

Chelsea

I returned to work this week and immediately noticed that somehow, since the time I submitted my PR, something had happened on master that meant there were now merge conflicts. When I initially submitted my PR there were no conflicts and all the tests passed. It was a bummer to say the least. Steph and I tracked it down to Toya having merged my branch with his, he made more changes to one of the files I had as well then submitted his own PR which was approved, squashed and merged before mine could be.

Unfortunately, the branch merge also caused a big problem on master because it did not include a migration file. I learned something VERY important today, migration files are not tracked, so once you make them, you have to manually add them to your commit. I was under the impression that when you did the make migrations an migrate commands that that was it. I had no idea that I had to add/commit/push those migrations. Now I do and I will not be responsible, albeit inadvertently, for messing up master again.

Friday, July 3

"Before we start," Arash declares, finally present for today's demo meeting, "I've been trying to set up the system for demos, but I'm getting an error when trying to connect to the backend." He asks if anyone else knows about this error and Yassin and Yazan try to help him resolve it by changing something in a constant file. Than mentions that sometimes when she makes migrations, the error gets resolved for her, although she doesn't use JupyterHub.

While the system builds again on his machine, Arash pulls up the project boards on the screen to review the done tasks for each team. Team 11 have finished fixing all their chat UI bugs in addition to their research for video chat changes based on breakout rooms. Team Rocket and A-Team, on the other hand, have a couple tasks remaining unfinished.

"Should we delete our done tasks after the meeting?" Than asks.

Shaking his head, Arash says he will archive them himself, and tries to identify which stories correspond to each task for both teams. He seems to have some trouble differentiating between lone tasks versus those associated with actual stories. He asks the mentors next if they want to demo anything. After a period of silence, he notes that Yousef was not able to make it to today's meeting.

The system finally finishes building. Arash proceeds with demos by creating courses and course offerings, uploading students and TAs to a Course Offering from CSV files, and uploading new avatars for users. After identifying some bugs and creating issues for them along the way, they spend the next half an hour once again discussing terminologies and consistencies in CSS styling due to everyone's different programming styles. Arash's efforts to reach a consensus on formatting guidelines end up in vain as Chelsea jokes that it won't be possible since everyone has a different opinion on what would be visually pleasing. Therefore, they conclude to have a story covering formatting guideline decisions for uniformity throughout the Ignite system.

Quickly after, however, the teams engage in another discussion regarding how emails should be saved in the backend for the instructor to copy and paste them into their Gmail or Outlook. They establish the next step as using those emails to create enrolments and then display a summary page of what was uploaded. After finishing up with creating all new issues brough up in their discussion, Arash decides to demo Team 11's stories for video chat changes based on breakout rooms in production mode during next week's demo meeting instead, since they are running out of time.

"We'll either extend our standup or have another meeting early next week to groom our stories for the next sprint." Arash concludes, promising that he will create new stories and define new RCs over the weekend. In the meantime, he delegates each team to work on bugs or stories that were created during today's demo meeting.

Chelsea

Friday was a rollercoaster. I started working on my bug fix the night before, but it wasn't going well. This morning Steph and I had a rubber ducky session and talked through both of our bug fixes then carried on separately to finish the code. I was able to have mine done, though I still have styling questions that I feel need to be asked in the Demo. Trying to get it approved was another matter. The code that I was working on was originally completed by Team 11 and approved weeks ago. Masoud wanted changes to that code base that were outside of my story. I did some of the minor changes, but the big ones turned into further bugs that I will likely work on next week. We had a successful demo of the story we started before the break. Apart from there being a problem with URL retrieval in production, it worked the way we wanted it to and looked good. I was able to demo the bug fix as well and bring up my formatting question. Ultimately, it's going to need to be decided and included in the Styling Story that Steph, Peter and Kevin are working on. Right now, there are so many different looking components that are essentially the same information and I think it looks sloppy.

Stephanie

The week off was well needed, but it was good to be back this week. It was a little hectic and unorganized with Arash not in the meeting on Tuesday, but we figured it all out. Our team accomplished a lot this week and it was nice to be able to demo a lot! Again, it was obvious that I'm better at working things out on my own as much as possible, then using Zoom meetings to help. We got some new bugs on Thursday and we managed to fix them all before demoing today! Front end work is still a work in progress and little bits of CSS will still be the death of me.

The demo session today was a bit frustrating with demoing a lot of things that didn't fully work and I think it took longer than necessary which means the planning part will be pushed to next week. It's hard to plan the week when we don't know what's up, but hopefully the meeting will be as quick as possible on Tuesday. Luckily our course is really starting on Monday so we can use Monday to work on that instead.

Chapter 20: Striving for Agile Practices

Tuesday, July 7

"If there is any one secret of success, it lies in the ability to get the other person's point of view and see things from that person's angle as well as from my own."

- Henry Ford

True to his promise, Arash has posted a to-do list for each team on Slack, based on the new stories and RC's he has created this past weekend. Likewise, today's standup meeting has been extended to accommodate for planning and grooming such newly created stories since last week's demo meeting went overtime.

"I'm not sure if Arash will be joining us to begin with." Mohammad tells everyone. After waiting in silence for a couple of minutes, he eventually asks each team to share their project boards on their own screens and give their updates.

Team 11 has successfully implemented their custom-made chat feature on JupyterHub and have added a button for creating breakout rooms during live lessons. In Team Rocket, Thomas and Peter are continuing to work on their bugs. Paul and Mihai will be looking at the new stories and picking some up today, having no updates to show off aside from testing **at the moment**. Kevin and Than have explored different options for implementing the data visualization platform and request to meet with Mohammad and Masoud later to discuss its requirements. In the A-Team, Bassem and Michael are continuing to fix bugs, link courses, and implement some changes based on Masoud's feedback. Although they need to pick up another task, they will conduct some research on their notebook-sharing

N. Raeesinejad et al., *The Ignite Project*,
https://doi.org/10.1007/978-981-19-4804-6_20

feature in the meantime. Chelsea, Stephanie, and Toya have merged all their bug fixes and will also need to pick something up from the new stories.

Mohammad thanks everyone for their updates and asks Yousef to take the lead on the grooming portion of the meeting.

"Can everyone see my screen?" Yousef asks, pulling up their repository backlog.

"Yeah." Mihai replies.

Yousef proceeds to open the first story in the backlog, which regards implementation of breakout rooms. "Okay, please read it by yourselves and size it in the chat."

A couple of minutes pass in silence before Mohammad and the team take it upon themselves to discuss the AC, although the depth of their discussion seems slightly more limited compared to previous discussions with Arash present.

"To be completely honest, I have no idea how to size this story since I never worked with Zoom before." Bassem admits.

"If you don't know, then don't size it." Yousef shrugs. This seems contrary to how the teams have been operating their estimation activities thus far; Based on some independent research that I have conducted myself over the past month, I was under the notion that it is a requirement for every team member to participate in the size estimation activity even if their size would be a "?" to engage in further discussions and clear any misconceptions. Perhaps this is a more theoretical practice that is not always executed in industry.

As a result of being told that they are not required to put down their size estimates if they don't know, the number of estimates is substantially reduced for each of the subsequent stories that are presented with limited explanations. I also notice that Yousef assigns the size label after at least around three to five members enter the same size each time. In my opinion, this further discourages participation in the estimation activity as well as submitting different estimates from the majority. During prior grooming sessions, Arash would wait until everyone had estimated and make sure the team defends and discusses any variations in their estimates. In the end, all of Team 11's stories are estimated as "5" without any objections.

A WORD FROM THE MENTORS

Ideally, we would have planned for each agile team to have their separate sprint planning, however, this was not possible due to time and resource constraints. Firstly, there was only one person as the acting scrum master who, also a product owner, did not have the capacity to facilitate multiple sprint planning meetings. Furthermore, due to the short 2-term duration of the Ignite project, we prioritized product over process, which is often the case for businesses. From the perspective of the Ignite team, deploying the MVP had higher priority than upholding perfect agile standards. While estimation and planning have their place in agile teams, they are only considered practices as opposed to principles or values. No matter how hard we may try to follow a specific process, we must sometimes pivot to ensure a deadline is met. In addition, due to tight deadlines, we could not allocate enough time to learning and practicing all aspects of the agile process, such as spikes. Another challenge we faced with following agile practices was working remotely, as we had no choice but to allow our developers to reveal their individual estimates in a live chat during sprint planning at the time, which resulted in more uncertainty within the team when finalizing requirements.

Since no one on the team had the technical context for the technology we wanted to implement i.e., Zoom for implementing live lessons on the Ignite platform, it was almost pointless to do story estimations when everyone sized story points to the max. In retrospection, we could have had the teams complete a spike and/or proof of concept to accurately estimate certain stories that involved new concepts and technologies; however, time restrictions would not have allowed for them to effectively execute this strategy. In the end, meeting the MVP deadline was made possible because of swarming and pair programming.

The next story is rather lengthy in description, involving the feature of instructors joining the lesson lobby, to be implemented by Team Rocket. This time, the chat fills up with "8"s.

"Do we need to break down the story?" Peter asks. "I remember Arash saying before that he didn't want stories with sizes bigger than five for our team."

Yousef purses his lips in thought before answering, "I think it should stay as an eight, since the maximum possible size is thirteen."

This engaged the entire team in another discussion regarding the complexity of the story further, although with limited clarifications given no one had a full picture of the system. They eventually decide to size a related story to get a better idea of their relative sizes. However, after another unanimous size of "8" for the next story, Mohammad tells Team

Rocket to start working on it with the assurance that they may revisit the story later to break it down further if needed.

"The new features are rather complex," Mohammad admits. "I will iron out the stories with Arash and Yousef this week to address everyone's questions." He also encourages the team to come up with their own ideas to pitch for some of the new features such as data visualization and gamification. Lastly, he mentions that the actual Ignite classes are officially starting again this week through Zoom and some other technologies.

Jerome

Yesterday we just worked on fixing some of the errors we encountered during last Friday's demos. Both were related to the sockets used for chatting. We only wrote some parts of the socket code on the ignite website, so we were able to get it mostly working after a while. Today we had a grooming meeting after the standup in the morning. As Arash wasn't there and he was the one who wrote most of the stories there was a lot of confusion as to what was going on and what we were supposed to do. Personally, I think it would have been a far more effective use of our time if Arash had been there to discuss the stories with us. After the meeting we worked on setting up the ability to switch between Zoom meetings based on a room you are assigned to.

Wednesday, July 8

Before initiating Team 11's standup, Mohammad mentions that Arash has posted some new notes on the Slack channel yesterday after discussing with him the details of the stories which were the objects of many team members' misconceptions during their last grooming meeting.

"Are you guys finished with your research on breakout rooms?" He asks.

"We're still working on getting the breakout rooms on the lesson panel and testing how it will work." Jerome says, "while we understand how to do it, it's just a matter of trial and error before we figure out the best way to implement it."

Seeing that Arash won't make it to this standup either, Mohammad asks Team 11 to go ahead with their updates, to which Yazan starts sharing his screen to display their project board.

Team 11 reports that yesterday, they worked on integrating the video extension into the lesson panel and will continue looking into it today. Then, they will test how breakout rooms are going to work. They will transfer the Zoom extension to the panel extension to reduce the amount of rework.

"When will you be done?"

"As long as we don't run into any major problems, we should be able to get that done today." Jerome answers.

Once they finish with their updates, Mihai initiates his onslaught of questions.

"If the video is not a separate widget anymore but in the panel widget, will there be a separate page just for the video within the panel?"

"It will be in the lesson portion of the panel." Yazan answers.

"Will the video be displayed with the lesson?"

"There will be a dropdown to show information about the lesson."

"If you click off the video, what will happen? Would you get booted out of the meeting?"

"I don't think you get kicked out," Yazan denies. "We're going to figure out the best way to make that work."

Mohammad chimes in, "The idea is that if you're in a classroom, none of the functionalities should be closed off."

Team 11 answers the rest of Mihai's detailed questions to the best of their abilities, revealing that some of their decisions need to be discussed and confirmed with Arash. Mihai, not seeming entirely satisfied, admits that the stories his team is working on this week depend greatly on how Team 11 decide to implement theirs. For now, they will wait on Team 11 to figure everything out first.

"Since Arash decided for us to use the Zoom chat instead of the one we had built before, it might take longer to implement those changes." Jerome warns. This is apparently news to Yousef and Mihai, seemingly unaware of this conclusion that was reached between Mohammad and Arash yesterday. Yassin points out that Arash did post this change in requirements on the Slack and reads it out for everyone in the call.

"For now, we will see if we can get away with just using the Zoom chat since the layout of having one extension will be simpler for finishing the MVP." Mohammad announces. "I'm looking forward to seeing what Team 11 produces by Friday."

Shortly after the meeting adjourns, I get a notification of a new post from Arash on Slack:

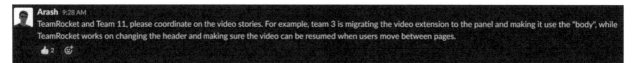

Thursday, July 9

Fortunately, Arash is present for today's standup meeting with all the teams.

Since yesterday, Team 11 has found an alternative platform to use in lieu of Zoom for live lessons, which they report to have found much easier to work with. After announcing they will start working on their stories for implementing breakout rooms today, Arash reminds them to regularly communicate with Team Rocket.

"We're planning on meeting with Team 11 right after this standup." Paul says.

Mihai adds, "In the meantime, we'll also pick up stories that don't depend on Team 11."

In addition, Thomas and Peter have merged their CSS bug fixes, while Than and Kevin are in the midst of finalizing all their stories for the data visualization dashboard.

"Do we need approval for the stories we have create for the data analytics dashboard?" Kevin asks and Arash reviews them quickly on their project board accompanied by their descriptions. Nodding in approval, he moves on to Team-A update.

Stephanie, Chelsea, and Toya have been researching and discussing gamification methods with Mohammad yesterday. Per his instruction, they have drawn out their ideas and will present them to him today for his feedback.

"Do you have a separate issue for this work?" Arash demands.

"It's encompassed by our other stories for defining achievements and tracking students' progress in the course offering." Stephanie explains. Arash nods, satisfied that all work is being tracked in the project board.

Lastly, Bassem and Michael have been troubleshooting to get Jupyterhub working and are now back on track and ready to demo their stories again from last week after doing some testing.

"I have a general question." Stephanie declares.

"Go ahead."

"How is the backend affected? Since everyone's stories are getting more complex, it's becoming kind of messy and difficult to keep track of them in the backend."

"If I may answer this," Than starts. "All the stories that Kevin and I are creating have a backend component, by adding new models or using existing models to record the data."

Arash nods in agreement before assuring her that A-Team's work is rather isolated compared to the overlap between that of Team 11 and Team Rocket.

"I have also added deadlines for every milestone." He reminds the teams.

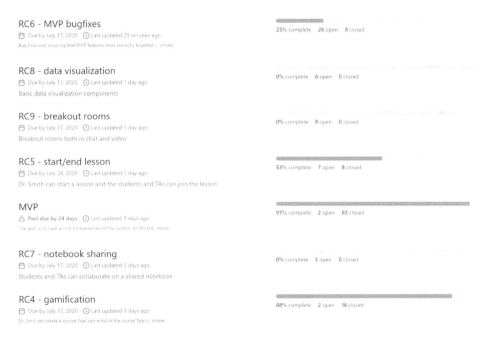

"I am under the impression that the teams have already discussed these deadlines during the latest grooming session."

"This is the first time we're hearing about it." Stephanie reveals with a concerned frown. Some other team members also seem perplexed to discover that the new deadlines for almost all their milestones are only eight days away.

Arash, with an air of nonchalance, says, "We can think about them for now and discuss whether we will be able to meet those deadlines during the demo meeting."

Jerome

The last few days we completely scrapped Zoom from the project. We did this because Zoom did not support everything we wanted to do with it. We found a new video streaming service that is much easier to work with. It would have been nice and would have saved us a lot of time if we had just started with this service. In the future it is probably important to do more research before choosing a certain service or library. I am a little worried that this service will not satisfy our demands but since we were in a rush to bring it back to where Zoom was we haven't really had a chance to look into it very much. This week we got the new service (Jitsi) working and we've set up a method to switch between breakout rooms in the client. One problem that has slowed us down this week is that another team is doing some of the UI and elements that are related to the video so we are stuck waiting for them to make progress.

Monday, July 13

While waiting for Mohammad to join Team 11's standup meeting, Mihai mentions to Arash that they were talking about switching to Jitsi as the new platform for streaming live lectures.

"I just want to make sure this is what we want to go with moving forward."

"From my understanding, Jitsi matches our needs." Arash says and prompts Team 11 for more context.

"Right now, we are using the Jitsi servers to run their service," Yassin explains. "We found that after looking at online discussion posts, the hard limit is seventy-five people per room and the soft limit is thirty-five, where afterwards the performance is affected. We were thinking of hosting our own servers to get around this issue to make it as scalable as we want."

"Peter has done some research on this well and found that hosting your own server is rather difficult." Mihai add worriedly and Yassin assures him that they have found an online source which provides examples on how to implement this.

Finally, Mohammad joins the meeting, but only for a brief moment to say he is caught up in back-to-back meetings before leaving the call. Starting their standup, Team 11 reports that they have worked on all the logic for the functionalities of the breakout rooms.

"Can I test your code in productions mode?" Arash asks.

"You would have to look into how JupyterHub would allow demoing both audio and demo." Yazan answers. "We may have to meet with Team Rocket again to coordinate this, since right now none of what we have implemented is only attached to a separate tab instead of the course offering page."

"In that case, I will demo your code in development mode on JupyterHub."

***

A couple hours later in the afternoon, I join the Zoom call again to attend the demo meeting that was pushed back from last Friday.

Having just met with Team 11, Mihai announces that Team Rocket's next step will be discussing what to get out of lesson pages moving forward. After reporting all the tasks that have been cleaned and testes in addition to newly discovered bugs, they spend a couple of minutes differentiating between stories which were demoed already from those that need to be demoed. It seems Team Rocket may have needed to clean up their project board prior to this meeting.

Similarly, A-Team will demo stories which couldn't be demoed last week, as well as some bug fixes.

"We have some minor bug fixes which have not been tested for quite a while." Bassem reveals.

"If you can't find someone to test your code within a day, let me know and I will hunt down people to test them." Arash says in a serious tone and Bassem laughs exasperatedly in response.

Next, they kick off the demos with Team 11's new video service that they recently discovered and implemented, with their assurance that all of its chat features have been finished.

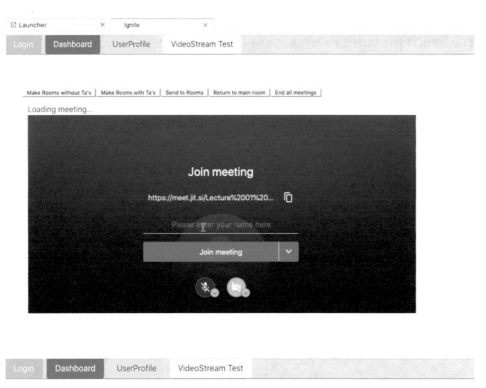

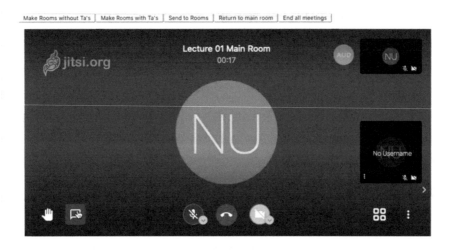

"Jupyterhub can't access the computer's video and audio, which is why it's not working in our video stream testing." Yazan points out. "However, the features for joining meetings, chatting, and being assigned to breakout rooms are working with the new Jitsi platform."

After quickly testing the chat feature on the platform, Arash announces that they will not need to worry about the stories for implementing their own chat anymore; they may revisit them later if needed. He then proceeds to demo Team Rocket's stories, which encompass some untested bug fixes. They also discover some new bugs along the way for which Arash creates new stories. Before A-Team can demo, he moves some of their stories to `Done` since they were confirmed while he was demoing the previous teams' stories. Afterwards, they briefly demo, on their own machines, how a user may upload avatars to their profile and share documents through a folder on the server.

With the demos done, they start their planning and grooming session.

"How do you guys feel about the coming sprint?" Arash asks A-Team.

"Well, Steph, Toya, and I are working on gamification and will have a meeting with the Ignite mentor team tomorrow to finalize our requirement definitions." Chelsea says.

Stephanie adds, "We will not be able to finish our stories for the deadline on Friday since we weren't able to start working on them last week."

Arash extends their deadline for the gamification milestone by one week after the team expresses sufficient confidence towards being able to finish those stories by then, given that their meeting goes well tomorrow. Arash tells them to notify him if they are blocked, not getting answers, or the requirements end up being too big.

"We also have enough work until Friday, with figuring out notebook sharing and moving its server on JupyterLab." Michael says and Bassem also expresses that the end of next week would be a more attainable deadline for completing their milestone.

"We still have a lot of outdated stories related to Zoom." Paul points out and Arash proceeds to spend the next couple of minutes updating their existing stories' AC and creating new stories to account for the new Jitsi platform.

"Team 11 and Team Rocket will need to continue communicating and working together this week as well." Arash reminds them. Due to uncertainties, Arash also extends the milestone deadline for breakout rooms by one week. Lastly, Than and Kevin also request for an extension of one week for implementing their data visualization platform.

"It makes sense to move deadlines around." Mohammad says, scratching his chin. "By when are we expecting to have an MVP?"

Arash sighs. "I'm hoping with these extended deadlines, the next demo should show that everything works and is deployable. All the functionalities should be there, even if they would have to fix small bugs and make the UI more presentable. There will be a lot of work for deploying the MVP. Hence, the 24th is now a hard deadline."

The meeting finally ends with Mohammad cancelling tomorrow's standup since everyone already knows what they will be working on.

Chapter 21: Backends, Frontends, and Merge Conflicts

Wednesday, July 15

"Always code as if the guy who ends up maintaining your code will be a violent psychopath who knows where you live."

- Martin Golding

There seem to be a lot more participants than usual in today's Team 11 standup; Even some members from Team-A are present. This may be attributed to the cancellation of yesterday's standup for all teams in addition to their hard deadline approaching.

Once Mohammad joins the meeting last, Team 11 begins with their standup.

"I have been working on getting the instructor to join breakout rooms." Jerome starts. "I finished the backend for it and a simple frontend, and now I want to combine my work with Yazan's frontend for the breakout room configuration which involves styling and UI." Similarly, Yassin and Yazan have moved onto more frontend-related tasks. With an end to their updates, the scope of discussion becomes open to involve the other teams.

Chelsea reports that the A-Team has been talking with Mohammad and the Ignite Club yesterday about incorporating badges for students on the Ignite platform. The challenge at hand is to integrate this feature with lessons and user profiles, which are currently being implemented by Team Rocket.

"So, you guys decided to use badges instead of points?" Peter asks.

"Yeah," Toya answers. "We won't integrate the point system used manually by the Ignite Club in the application unless needed."

"The Ignite Club had developed this point system themselves in the past." Mohammad clarifies.

"These were mostly based on end-of-session and social media challenges." Chelsea adds. "For now, we decided to keep the point system manual, since many of its functionalities - like dropbox for project submission and the grading system - are out of our scope at the time being." Nevertheless, it seems Mohammad will allow the Ignite Club to continue these practices offline for the time being. Lastly, they pitch their idea to include certifications with different criteria, on which Arash agrees and suggests that their grading system could be automated with Jupyter's `nbgrader` extension.

"The reason why I asked if A-Team will use badges is because when a lesson ends, we made it so that something pops up and think it could be valuable to include the badges gained from the lesson that the student just attended." Peter says.

"The student can only receive up to one badge." Stephanie points out, suggesting that they could be redirected to view all their badges on their profiles instead.

Mohammad nods in agreement. "You could also add a component to allow for students to share their badges on social media."

"Well, we're not too sure of the complexity of adding this component." Peter admits.

"It can be something to think about down the road."

It seems all the teams are much more engaged in discussions about the system features when it comes to the UI and UX aspects.

A WORD FROM THE MENTORS

It is important to distinguish the difference between User Interface (UI) and User Experience (UX) when it comes to designing and developing software products, applications, or services with user interactions. Human-first approaches and layouts are used in conjunction throughout the design process with the user in mind. Below are key questions that UI and UX designers address respectively and questions specific to the Ignite project.

	User Interface (UI)	User Experience (UX)
Key Question	How does the user interact with the product?	What is the overall experience of the user?
Examples in the Ignite project	How will instructors join breakout rooms? Which elements should be included in the popup after a lesson ends?	How easy is it for students to navigate to their earned badges for lessons? How does earning badges or points make students feel?

Stephanie

Yesterday we met with the Ignite Club to see how they are currently dealing with gamification, and how we can implement ours into the system. They are using a manual process for the point system and are grading the challenges which isn't something we can't do for the system yet. Chatting with them and Mohammad, we're going to implement the badge system into the platform while they keep the point system manually for now.

Today, Chelsea, Toya and I met to break up gamification into the correct stories. Chelsea and Toya are going to focus on the backend, which we think is going to be like Avatars and it's great that we can scale all the work we did for that. I'm going to focus on the frontend for the profile and maybe the other parts.

After looking at the user profiles, the student profile isn't broken out as much as we thought, so our original plan for the format of the page had to be changed. I'm going to focus on making the badges work, and then revamping the user profile page in general to look more like other profile pages like GitHub, LinkedIn, etc.

So far today I've learned that working with the frontend is frustrating haha. Things that I think would work for layout don't always work, but I know that the more I do it the more I'll learn so I'm glad I'm taking these stories!

Thursday, July 16

Since yesterday, Team 11 has finished the fronted for the breakout room configuration and combined all their work which is waiting to be approved by the mentors. These functionalities will be combined with the video-streaming page, which Team Rocket is currently in the process of finalizing. In the meantime, Kevin and Than have set up a separate widget for their data analytics extension and have started working on the UI for challenge and course pages.

In the A-Team, Bassem and Michael are working on a button for the Jupyter notebook.

"How did you guys figure out getting TA information from the backend?" Bassem asks Team 11.

Along with Arash's input, they explain their process towards remembering the mapping between TAs and students in the database.

"We can provide something like a REST API extension for Bassem and Michael to have easy access to the TAs in each breakout room." Yazan offers.

"We may not need to create another API... but we will discuss this further." Arash assures.

Lastly, the A-Team is continuing to implement the gamification features of the system I.e., creating, assigning, and displaying badges. In addition to completing these tasks, Stephanie requests to redo the user profile since they did not account for it to be separate from the student profile in their UI mock-up, which was already approved by Mohammad, to which Arash also nods in approval.

"On the side, I've been looking into static file serving." Toya mentions.

"I suggest you look into an online document called *WhiteNoise* for that." Arash says and with no further comments, he ends the meeting, lastly expressing that he is looking forward to their demos tomorrow.

Toya

The issue in production is that the Django server will not serve static or media files in production unless you override the source code. This means gamification and basic profile won't be complete until production environment is set up. I looked at options for deploying in production. Firstly, I found that WhiteNoise is not suitable for serving user-uploaded "media" files. For one thing, it only checks for static files at startup and so files added after the app starts won't be seen. More importantly though, serving user-uploaded files from the same domain as your main application is a security risk.

In addition, using the local disk to store and serve your user media makes it harder to scale your application across multiple machines. For all these security and scalability reasons, WhiteNoise proved to not be a viable option. It would be much better to store files on a separate dedicated storage service and serve them to users from there. The Django-storages library provides many options like Amazon S3, Azure Storage, and Rackspace CloudFiles.

My top 2 choices would be AWS (Amazon S3) or Digital Ocean. Amazon S3 is larger scale and has 5GB storage in their free tier, which is more than enough since uploads are limited to maximum 8KB. Digital Ocean is also a good candidate, but just smaller in scale than AWS More. I investigated top cloud services as well as how to set up Amazon S3 and Django through online tutorials and official documentation on Django-storages, which also contains sections on Amazon S3, Google Cloud Storage, Azure Storage, and Digital Ocean.

Friday, July 17

10 minutes pass in silence while the team waits for Arash, Mohmmad, or one of the mentors to join the meeting.

"Maybe Arash and Mohammad aren't on the call because they know that we have nothing to demo." Mihai jokes, breaking the silence.

Chelsea laughs. "I could definitely handle being forgotten about today!" Some other members nod their heads while others chuckle.

"I was talking with Arash just an hour ago and he was trying to boot everything on his system for the demo meeting." Bassem reveals, shrugging.

After a couple more minutes, Mohammad finally joins the call with an excited grin.

"Does everyone have exciting demos to show today?" I see Stephanie grimacing while shaking her head and Chelsea maintaining a nervous smile. Fortunately, Arash also joins the call before anyone can provide a lackluster response and immediately starts reviewing all the team project boards.

While Team 11 and Team Rocket have successfully integrated their work together on JupyterHub, they will not be able to demo their video, audio, and screen share functionalities due to some issue encountered with HTTPS. Aside from combining their code, Paul and Mihai are still restructuring how lessons work in the background, thus having nothing to demo either. Similarly, Than and Kevin only have some frontend UI done which is not connected to their backend. Although these would not technically count as "demo-able" features, Arash urges them to show how this looks from their screens for the rest of the team.

"A lot of our tasks must be complete prior to being able to be demoed together." Bassen notes.

Arash hums. "Is there anything A-Team can demo?"

"Not really." Stephanie denies. "Our badge feature is also not fully done. Should we wait until Team Rocket is done merging their code before changing our backend?"

He nods and tells Team Rocket to prioritize finishing their backend for the A-Team.

"There's a lot to add to the backend." Than admits, suggesting breaking down their backend into more fine-grained models and Arash agrees wholeheartedly.

"Bassem and I also ran into a roadblock with breakout rooms," Michael adds. "We were experiencing some serious challenges with testing it."

"In that case, it's better to demo Team 11's work and focus on breakout rooms." Arash says. "Then, we can discuss changes being made to the backend so that everyone is on the same page when continuing to work on these tasks next week."

He spends the next couple of minutes building the master on his machine. After running into some internal server errors, the site finally loads, and he asks Team 11 to guide him through their demo. Per Yazan's instructions, he starts off by creating a course along with a course offering and uploads a CSV file of students and TAs.

On the topic of enrollment, Bassem and Michael ask Arash whether they can add a single student to a course offering with their name and email.

"It's not needed in the production case, but useful for the test case."

"The reason we're asking this is in the case of a student needing to enroll for a class halfway through the semester." Bassem reveals.

Arash ponders for a few seconds before saying that this is not a requirement at the moment, since they are building the system based on the requirement that all students are enrolled together by uploading one CSV file to the site from the very beginning.

"Can an instructor's role be toggled, seeing that it's possible at the moment on the website?" Toya asks.

"It's a good idea to promote a TA to teach and give control of the lesson to someone." Mohammad says.

"This would similarly be possible by simply having someone share their screen." Arash points out. "Toggling the role of an instructor should not be possible." He creates a new story for removing this feature after Mohammad's voices his approval.

Next, Yazan tells Arash to create a breakout room with TAs and students, which can be moved around rooms.

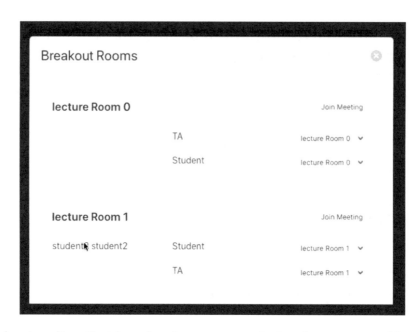

"It was my understanding that breakout rooms needed to be approved by the instructor." Arash says inquisitively.

Mohammad explains, "You know how in a classroom I tell everyone to work on a problem for 10 minutes in little groups and afterwards tell them to stop? I should have the ability to send people to breakout rooms and bring them back."

Nodding, Arash adds that it would be good to have some sort of confirmation after assigning people to breakout rooms when logged in as the instructor.

"Currently, this is a placeholder UI, and it will be much more user friendly once combined with Team Rocket's code." Jerome explains assuredly.

"It would also make our work a lot easier if Team 11," Bassem adds, "since they created the backend API for breakout rooms and could send the information that we need through an API call."

Arash tells them that if their backend is ready, they can submit it without the frontend. However, a couple of other members note that this will not work with their new frontend changes. It seems everyone's backend changes are held up for now.

Leaving this aside as a separate matter to be addressed later, Arash prompts Than and Kevin to share their data analytics progress. Kevin starts sharing his screen and introduces their analytics dashboard for which they have been creating components containing different charts and graphs.

"Thanks guys!" Than smiles and notes that there are still a lot of details to figure out and implement later to the analytics.

"I think there are a lot of options we can think about." Mohammad comments and suggests for Than and Kevin to look at how data analytics are presented on YouTube. "It's all about filtering data and comparing data based on different demographics."

"It would be nice if it were more customizable as well." Arash adds but quickly points out that these functionality comments should be considered as low priority, and that he doesn't want to change the scope of their work.

The teams proceed to spend the next couple of minutes discussing their backend changes – like adding endpoints for breakout rooms and lessons – in order to prevent merge conflicts in the future.

"By the way, can we have another project board for Team Rocket?" Mihai requests, "Our work is diverging and it's becoming harder to track all our tasks."

"Sure." Arash creates a new project board by selecting an automated Kanban with reviews. "What do you guys want to call it?"

"Team Sprocket." He answers with a broad grin. Chuckling, Arash enters the name of the new board as 🗂 `Ignite-Team-Sprocket`.

"We can move the cards over from the original board in our own time." Paul promises and Arash reminds them to match the automation settings on the columns to their original project board as well.

Lastly, he announces that he will figure out deployment on cloud with Peter, who has done preliminary research on it, as well as Toya, who is interested in being involved in deployment.

Peter

This week, my group has been focusing on creating the Lesson "rooms" containing the page holding the Jitsi video component for each lesson. My group met with Team 11 multiple times to gain an understanding of the various micro services Team 11 had built to support video streaming. As is the nature of Agile development, there are many parts of the source code I was not involved in designing nor writing so this was a good opportunity to expand my scope of familiarity of the source code. Since these meetings with Team 11 were held with my entire team of 4, these meetings were essentially impromptu code reviews as my team was often asking questions and requesting changes to better suit our needs.

Yesterday, my group met once again to review Team 11's backend Lesson Video services with the objective of confirming whether the backend currently supports all the services required by the frontend portion. Resulting from this meeting, a few major gaps were identified. My team then devised a high-level major backend design change that required Team 11's feedback before we moved forward with implementation of the design change.

After meeting with Team 11, we determined that the major design change was necessary and would be more efficient if Team 11 worked on the implementation since they were out of work for the remainder of the sprint and already familiar with the existing code. Since the major design change to the backend would no longer be compatible with the frontend UI, it was determined that the best git workflow would be to implement both the backend and frontend changes on a single branch (i.e. both Team 11 and Team Rocket would be working on the same git branch). Throughout the entire day, our teams worked on implementing our changes to the branch while participating in a collective group video call. Oftentimes, our teams would have to get feedback from the other to ensure our designs would ultimately be compatible with the other team's implementation. This work session was extremely productive but introduced some major challenges.

Since changes to the backend were in development and the previous UI had dependencies on the previous backend, development of the frontend portion was challenging. As we were developing the frontend, we could not practice the usual development workflow of "make a small change, then test it visually on the UI to confirm it works as intended". Instead, we were forced into a workflow that required us to develop the code without being able to confirm functionality until the very end. This workflow forced us to spend significantly more time to stop and think through our designs before implementing, and again after small code changes were made.

Chapter 22: Design Debates

Tuesday, July 21

"If you want your users to fall in love with your design, fall in love with your users."

— *Dana Chisnel*

Since the last demo meeting, Team 11 have been working on refactoring the backend based on their discussions on Friday and will continue working with Team Sprocket to connect to the frontend. In the meantime, Team Sprocket i.e., Paul and Mihai have been refactoring their code which is yet to be merged with that of Team 11 in order to have a fully functional lesson model. In Team Rocket, Kevin and Than have nearly finished implementing the data analytics front page and will start working on the backend and integration.

"We have some questions regarding whether a model is being built for challenges." Than says.

"I'm not sure if anyone is doing the frontend for challenges." Arash rubs his chin. "It should be done by either Team Sprocket or you two."

"Well, it will need file sharing and breakout rooms, since challenges are coded by the students." She explains.

"We will also need to discuss our data collection surveys with Mohammad later." Kevin adds and Mohammad immediately voices his agreement with enthusiasm.

In the A-Team, Bassem and Michael are currently waiting on the backend merge and will work on implementing the TA's perspective of student notebooks once complete. Steph will continue working on organizing the user profile while Chelsea is finishing up on adding badges to the backend. Lastly, Toya will be setting up an Amazon test environment for production.

With the standup part done, Arash makes his next announcement to the team. "I have been talking with a UI designer to give us a mock design to have a better understanding of the look and feel of the website." He proceeds to ask for the team's opinion on the first mock design:

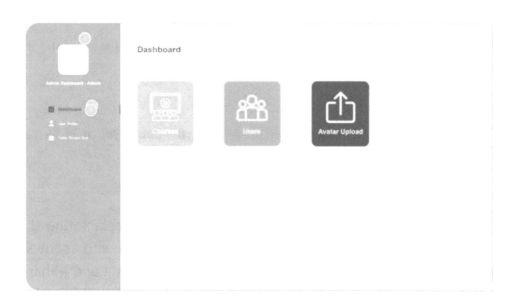

"It looks clean..." Bassem starts.

Stephanie immediately nods. "I love the color palette!"

"... but I'd have to disagree on that." he finishes.

"Yeah, these colors seem dated." Peter adds. "White with minimal colors would look better."

"There is obviously a dichotomy between what the male and female members think look good." Chelsea shrugs.

"I'm just saying the colors aren't appropriate for a learning application!" Bassem defends.

Peter nods in agreement. "It looks more like a personal application instead. I think we should follow Google's example for simplistic designs."

Before they can debate further, Mohammad and Arash quickly jump in to assure that everyone's opinions matter. Mohammad notes that the design looks modern enough and urges the team to move on with discussing other elements, such as the desktop/mobile views and dark/light themes.

"We should follow what the end users say instead, so that we won't have so many clashes of opinion within our team." Toya suggests.

Arash coughs. "This discussion with the Ignite mentor representatives would have to be soon, since we need to submit what we want for the UI designer to have a design ready for us within two to three weeks."

"In our previous chats with the mentors, they did say that they wanted to have their Ignite logos in the application." Stephanie reveals.

"That's good, just keep in mind one thing." Mohammad says. "Even though the Ignite mentor representatives are the end users, they are still a revolving team with new club members every year. My opinion would be prioritized next as a stakeholder, but I don't really have a strong opinion on design..."

"We can reach out to a graphic designer within the Ignite mentor team to help us better integrate the Ignite color scheme and logos in the UI design." Arash suggests.

"Would it be a good idea to have a small team of two to three people meet with the Ignite folks?"

"That's what I was proposing." Toya insists. "We should have the people on the styling team chat with them; so, that would be Steph, Peter and Kevin."

However, Stephanie and Kevin are quick to voice their concern that they do not have the time to meet this week, swamped with completing their own stories. Fortunately, Arash does not intend for the design to be implemented now, but for the team to simply decide on the color palette and general form of the site to hand off to their designers. Before ending the meeting, Mohammad requests whoever else with time to also be present in the discussion with the Ignite mentor representatives.

A WORD FROM THE MENTORS

In product design, it is important to distinguish between designing for yourself versus designing for your primary end users. The former prevents you from understanding the bigger picture of your product and limits your design and development scope. When you regard yourself as the perfect end user, your design will not support the varying needs and goals of your actual users.

Requirements Engineering (RE) is the process of defining, documenting, and maintaining requirements in the design phase of software systems. During this process, verification and validation serve crucial roles in ensuring the software meets its specifications and stakeholders' problems domains were understood correctly, respectively. In other words, we must address the following questions:

- ✓ Did we build the system right?
- ✓ Are we building the right system?

Certainly, the concepts of verification and validation in system and software engineering also applies to UX design with different testing techniques, such as usability tests, UX health checks, and surveys.

Wednesday, July 22

Arash is not present for today's standup, so Mohammad asks someone from Team 11 to share their screen and show the project board instead. Yazan reports first that aside from working with Team Sprocket to get their functionality working and connecting their backend and frontend, he has been helping Toya with getting his production mode to work.

With Arash's absence, Mohammad encourages Team 11 to ask him their technical questions. Yassin then proceeds to share his screen and show the UI after building his backend.

"When an instructor clicks on a button to create breakout rooms and assign students to TAs randomly, would you like to view a list of the rooms that the students are assigned to or are currently in?"

"I had discussed this in great length with Arash before and I want to stay consistent with what we had decided." Mohammad reveals. "I want to be shown a map of everybody in the rooms and be able to jump in any room."

"Are you asking to show a student's assigned breakout rooms even if that student is not in a lesson?" Mihai asks Yassin. He falters for a moment, so Mohammad continues his clarification.

"What I have in mind is a global map of all the breakout rooms, just like in Zoom breakout rooms."

"So, should they be assigned based on people who are in the current lesson or just the whole course?" Yassin asks.

"You end up merging the groups." Mohammad finally tells them to not worry about this matter for now, since instructors may manage breakout rooms manually if some students didn't show up to their lesson.

Satisfied with his response, Team 11 and Mihai thank Mohammad and they end their standup.

Friday, July 24

Today is the revised official MVP deadline. I'm rather excited to see what the whole team has accomplished.

"I've been holding off on shaving my beard until the MVP!" I hear Mohammad announce excitedly as soon as I get connected to the call.

"You may have to wait a little bit longer." Stephanie says with a sheepish smile.

"If it's just enough, I guess I will resort to shaving it." He chuckles.

Clearing his throat, Arash starts sharing his screen and presents the current progress for each team:

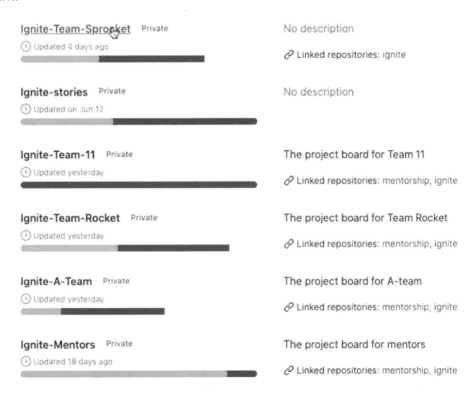

As shown, Team 11 have finished all their tasks for this week in close collaboration with Team Sprocket. Since the Data Analytics backend is not done yet, Kevin and Than from Team Rocket do not have anything to demo today. Nevertheless, Arash encourages them to present their current progress on their own machines.

In the A-Team, Bassem and Michael couldn't finish solving some sort of injection security risks. Meanwhile, Stephanie, Chelsea, and Toya have finished implementing badges and updating the user profiles. However, the badges may currently only be added, not displayed on the site. After building the master branch on his machine, Arash presents their gamification feature, where the admin may upload new badges from their dashboard:

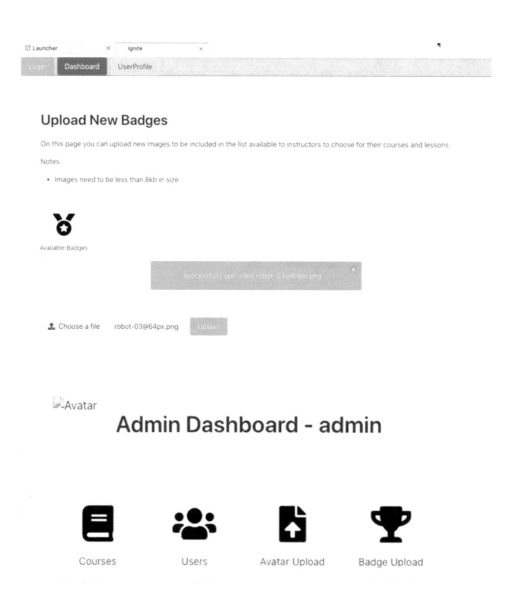

Next, Team Rocket proceeds to demo their work on their own computers to show a 3-person Jitsi call. After clarifying some specifics of live lessons with Arash and Mohammad, Peter starts sharing his screen to demo a Jitsi group call between Mihai, Paul, and himself:

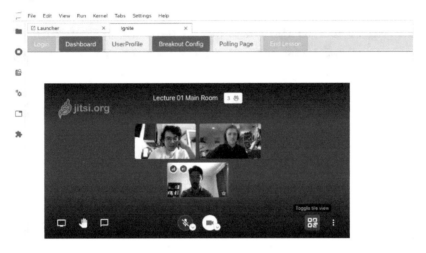

They then show how to assign breakout rooms. Instructors can now split students into separate breakout rooms and customize the maximum number of rooms. In addition, students may now be assigned to different rooms which the instructor can join at any time.

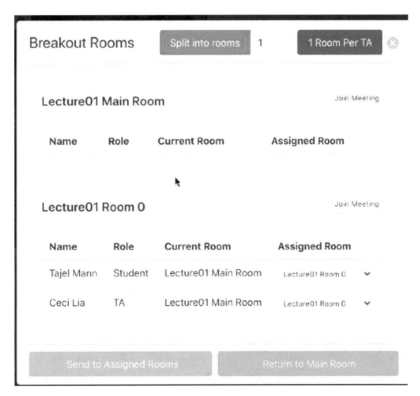

"Can you chat in the breakout room?" Arash asks.

"Yep." Mihai nods and proceeds to use the Jitsi chat feature to talk send messages with Paul and Peter.

He then directs them to exit to the main live stream room and return to the breakout room. When they return, the breakout room chat seems to no longer show their user IDs from before.

"The chat history is maintained if you leave and come back but the ID would be different." Paul explains.

"Would it be easier to just have the students stuck in the breakout room?" Mohammad asks.

"We could, but this situation would repeat for the main lesson room as well."

Arash then requests Mihai to share his screen to demo the TA view, in which he shows that a TA may view the notebook summary and hide videos in a tab. He then asks Paul to demo the live stream from the student's view. His shared screen shows there is now a finished task button and help button for the student to ask questions from the instructor and announce that they have finished their coding tasks.

"Do you want the help button here or in the notebook?" Arash asks Mohammad.

"It would make more sense to put it in the notebook."

"Jitsi does have a raise-hand feature." Paul notes. "Although one concern is that if someone keeps clicking on the raise-hand button, the TA would be spammed with notifications."

"And we don't want other students to get notified if one student has a question, it can be a distraction." Mohammad adds, rubbing his beard.

Paul nods and continues explaining, "The only way they could take care of that is through a private chat with the TA. We will have to take care of this probable situation."

Lastly, they demo how a TA can bring all the students back to the main room.

"By the way, does the instructor need to configure anything with Jitsi to do all of these things with live streaming?" Arash asks.

"They have already been taken care of in the backend." Yazan assures.

"We also had a live button beside a lesson in a lesson list to show if a lesson is currently live, but this functionality was merged after you pulled the code to your machine, so it is not visible in this demo." Mihai explains.

Lastly, they discuss the prospect of additional features, such as the history of live lessons and their durations. Mohammad notes that this data would be useful for other teams' implementations, such as gamification and data analytics.

Arash praises them for a good job and moves on to Team 11's demo of how to use the Amazon server. Yazan briefly guides him through video streaming from the admin side using the server. Next week, Team 11 will figure out how to scale this to work for the number of users that they need. Arash agrees that their main priorities going forward should be deployment, security, and scaling.

He then demos Than and Kevin's progress for the data analytics frontend, during which Kevin explains all the different ways an instructor can view and filter the data pertinent to their lessons.

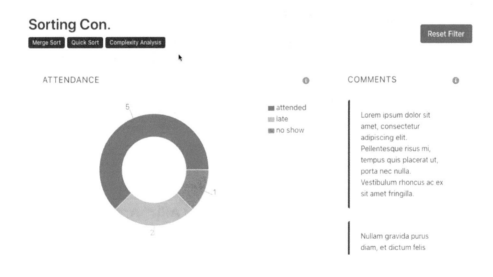

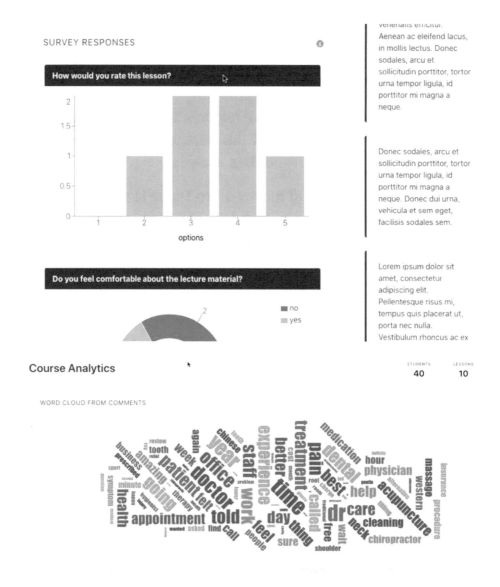

"How easy would it be to access analytics and the Jitsi API?" Arash asks Team Rocket and Team 11.

"We would have to look into the API." Yassin answers. "It may be easier to have our own raise-hand button and create a queue for how many hands were raised during a lesson."

"There may also be another metric involved with this, such as the amount of time spent during filesharing between a TA and a student." Bassem adds.

Kevin then mentions that they have also created a lesson list:

LESSONS

#	Title	Date	Attended	Late
1	Introduction to Programming	2020-5-1	40	0
2	Python Basics	2020-5-8	34	3
3	Conditionals	2020-5-15	30	0
4	Nested Conditionals	2020-5-22	25	5
5	Loops: For, While, Do-While	2020-5-29	25	2
6	Nested Loops	2020-6-6	30	2
7	Arrays	2020-6-13	33	0
8	Objects	2020-6-20	30	3
9	Classes	2020-6-27	29	1
10	Review	2020-7-3	40	0

"I like the look of everything." Yazan compliments and they thank him.

"We also wanted to demo a survey which is not merged yet." Than reveals. "We have a default survey to be displayed at the end of every lesson and this can be modified by the instructor".

"Maybe you can have a different number of options in the answers." Arash suggests.

"The standard number of five answers is good enough for now." Mohammad says. "You can have what Arash suggested as a low priority task for the week."

Lastly, Than shows everyone the different backend models that that she has designed with Kevin over the past week in order to handle all their collected data after discussing the layout of the survey questions with Mohammad.

In the A-Team, Bassem and Michael reveal they can't demo today because they needed a core functionality to be done by another team which got merged just today but will instead have a five-minute demo ready for the next standup meeting. The gamification features will also be demoed next week.

"Okay, starting on Monday, we will switch gears from actively adding features to making existing features more robust and bug-free." Arash announces. "I will work with Team 11 and Peter to set up some milestones for deployment. Meanwhile, Team Rocket wants to add some quality-of-life changes to the frontend. Lastly, I relayed your feedback to the UX designer since last time and we now have a new mock-up."

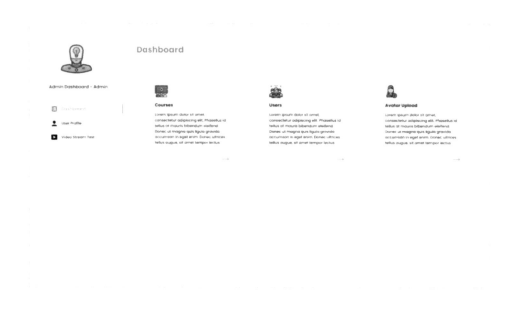

Most of the teams have more positive reactions to this design compared to the last one, commenting that it looks simple and clean.

"Thanks for mentioning the Google educational site layout Peter." Mohammad says.

"For sure, I do think the page could have some more personality, since it looks kind of too white." Paul notes.

"The Ignite team did tell me that they want more color as well." Stephanie adds.

On that note, the team spends the last couple of minutes of their meeting brainstorming areas to add more color across the website.

After Arash summarizes their general plan for next week one last time, Mohammad expresses his gratitude to the team for their time and great work. The meeting ends on a positive and determined note.

Mihai

This week, our group continued working on our user story of getting the video streaming working. We had to rework some of the frontend and worked closely with Team 11 to hook into their revised backend. There weren't really any surprises this week - it was just a matter of working on it and finishing it prior to the deadline. On Thursday, we tested things out and rehearsed for the demo on Friday. We found some minor bugs and improvements and Team 11 was able to finish them early Friday morning (in time for the demo).

The demo went well and this coming week, we plan on working on a number of improvements (quality of life, robustness, user-friendliness, etc...) to the video streaming feature. One thing I learned this week was that VSCode Liveshare lets you share your localhost port!

Section VI. Post-MVP

Chapter 23: Deployment

Thursday, July 30

"The most important tool we have as developers is automation"

- Scott Hanselman

Tapping my pen against my notebook, I wait in silence with the rest of the meeting participants for Arash or Mohammad to join today's standup. Since Yousef is not present either to direct the flow of the meeting, I wonder if anyone will initiate the standup at all. Dropping my gaze from the Master students' yawning faces, I decide to review my notes from this week's past meetings in the meantime.

With deployment right around the corner, the Master student teams have switched gears from actively adding new features to making existing features more robust and bug-free. Team Sprocket has taken the initiative to add quality-of-life changes to the front-end at their own discretion. After going through many model designs, Team Rocket is now finalizing the backend model for their data analytics feature, which spurred close communication with other teams on how to access certain data from the pages for which they are responsible. Meanwhile, the A-Team members are continuing to work together with Yousef towards finalizing the live document sharing feature between students and TAs, while Team A+ are continuing to develop the gamification feature through badges on student user profiles.

As a result of Arash's expectation that this week's demos will be running on a server in lieu of Docker, Team 11's current main priorities are deployment, security, and scaling. The undergraduate team has been working diligently on connecting Jitsi to AWS[34]; they were first required to deploy Jitsi as well as a single plain Django project to AWS, which

---

[34] AWS (Amazon Web Services) is a public cloud computing platform provided by Amazon to deliver computing services via Internet. This secure service provides additional functionality designed to help your business grow without the costs of creating a true private cloud.
Amazon Simple Storage Service (S3) is a service offered by Amazon Web Services (AWS) which provides object storage through a web service interface. The basic storage units of Amazon S3 are objects which are organized into buckets.

N. Raeesinejad et al., *The Ignite Project*,
https://doi.org/10.1007/978-981-19-4804-6_23

will later be used to run the entire Ignite application. Today, their focus is on troubleshooting to access the database and get their Django project backend running on AWS without hard coding. Lastly, they have recently created a file called "Instructions on setting up AWS Elastic Beanstalk" for other teams to follow in order to deploy their own environments.

"Hey, Team Eleven, why are you guys sometimes called Team 3 by the mentors?" Chelsea's question pulls me from my trance of thoughts.

"Actually, our team's name is 'team one one' which is binary for three." Yazan explains and Chelsea looks like she just got an epiphany. Another moment of silence passes, and I eventually sigh and unmute my mic to suggest someone to message the mentors on Slack. After sending private messages to both Arash and Mohammad, we resume waiting for one of them to reply. Following a couple of minutes comprised of the undergraduate and Master students chatting about their courses, Arash finally joins the call, apologizing for having overslept.

"Team 11, please go ahead with your update." he says while adjusting his glasses.

"We have figured out how to get the backend on the server." Jerome says confidently. "We're planning on deploying our actual Django proj-."

"Sorry I'm late guys!" Mohammad suddenly announces, finally also present in the meeting.

"No worries, we're just done Team 11's standup." Arash assures. "Does anyone else have any updates or questions?"

Bassem clears his throat. "Michael and I have been chatting with Masoud about the difficulties in accessing information, and we are just wondering if our relational database will change."

"It shouldn't, unless there is good reason for it to be changed."

"We just noticed that there seem to be chained requests or API calls in several points across the frontend."

"Do we need to have continuous requests to refresh something? It seems unnecessary." Toya notes.

"We don't need to do that." Arash agrees. "What we should do instead is reduce the number of timed requests as much as possible. For usability improvement, we can remove them, unless they're required for functionality. We should also remove the button for refreshing the frontend since it could be spammed and is not really needed."

Next, Arash shows the updated UI mock-up designs to the team, including both a web and mobile view of the platform.

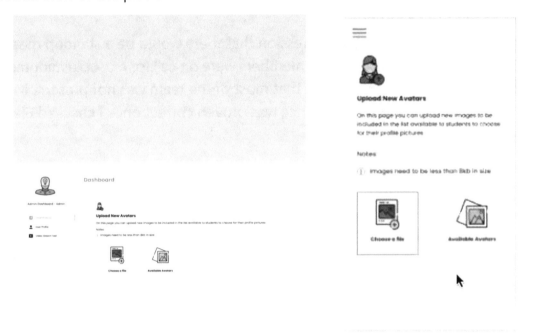

"I want to dedicate a portion of time to restructuring the UI in the next sprint." he mentions and with one last apology for arriving late, he ends the meeting.

Jerome

The last two days we have worked on deploying Jitsi and the Django backend to AWS. As none of us are very familiar with the process we've had to do a decent amount of research to figure it out. There is some documentation but not as much as we had hoped so we've had to rely a lot on trial and error. Many of the things we're doing, such as the certificates that are used for security, require the administrator AWS account to create. As we do not have one, we've had to ask Arash to make them for us and then send them over. So far, we have been successful with Jitsi and have set up a basic Django backend so tomorrow we will try with the actual Django backend.

Tuesday, August 4

The next morning, I was under the impression that there would be a standup meeting for all teams. Aside from myself, two other members were on call for a couple minutes at 8:30 am but quickly left the meeting, seeing that most of the team was not present. My hunch that the standup was cancelled in advance was proven correct once I checked Slack soon afterwards.

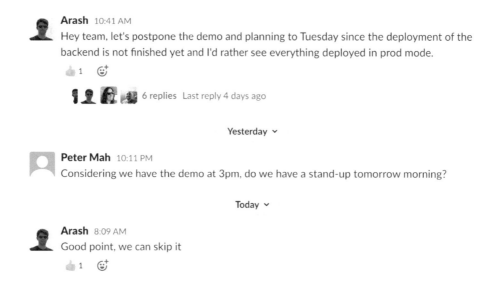

3:00 p.m.

"Please prep your systems to demo on your own machines today; I have unfortunately run into an error when trying to demo myself." Arash tells the team. In the meanwhile, he reviews each of their project boards and asks whether they can demo.

Although Team Sprocket has no done tasks and no external parties tested them since they made minor changes to the backend, they mention they can still demo what is functional on their machines with testing in progress.

Team 11 just had their code accepted and they will demo their AWS backend and Jitsi on their own DNS[35].

"By the way, when you squash and merge, you will need to click off for debug." Yazan notes to the team and Arash tells Team 11 to include all these instructions in their demo

---

[35] Domain Name System (DNS) translates human readable domain names (e.g., www.ucalgary.ca) to machine readable IP addresses (e.g., 136.159.96.125).

as well as frequently update their documentation with all configurations and changes to the server so that the rest of the team may learn to reproduce on their own.

In Team Rocket, Than explains that although essentially all their APIs are working, there was a lot of overriding involved for data gathering.

"Can we get one of the metrics needed for data analysis from Bassem and Michael's notebook?" She asks.

"Michael, Yousef and I have actually been working this out." Bassem explains. "We created fields that contain details based on their filesharing mechanism. You should be able to add your own fields depending on what data you need once it's merged."

"We may need to change the frontend to reflect those changes. It'll be difficult for us to generate data while trying to capture as many bugs as possible since data can't be guaranteed to be generated in the same order that we activate our surveys throughout the lifetime of the system." Than grimaces. "For the moment, we have one default survey that is customizable and are still figuring out how to display it."

"Will you two be demoing anything today?" Arash asks them.

"I could demo quickly on my machine, but there were not many changes since last time, since we mostly just connected the frontend to the backend." Kevin says and Arash accepts that since they already demoed UI functionality, they may be skipped today.

"We don't have much to demo either." Bassem says. "But we did fix a lot of bugs and are waiting for a review from you." Arash sighs but smiles again in relief when Chelsea reveals next that Team A+ can demo badges.

Team Sprocket kicks off the demos with showing their quality-of-life changes to the live streaming UI, including resizing the streaming video and adding some buttons for lesson description, leaving breakout rooms, and returning to the main page. They also explain the added restrictions and permissions for each type of user in the system. Next, Team A+ show how badges will be displayed on a student user profile after they merge with master along with how avatars look permanently on the user dashboard.

After cleaning up their issues in their repository, Arash notes that they will need to add some more bugs to the list and then have Mohammad let them know what they need to prioritize for the coming sprint.

"Going forward, deployment has top priority." Arash announces.

Mohammad nods. "Also, UI changes to match the new design and notifying TAs when a student has a notebook question both have precedence over recording and replaying lessons. Securing the Jitsi endpoint such that only our app can create meetings is also prioritized. Meanwhile, tracking students' progress in course offerings, data analytics, and other course UI stories are still in progress."

Deploying Jupyterhub on the Google Cloud Platform will require all of Team 11 and Peter. Apparently, there is a lot of background information required, so they agree on allocating ten days instead of one sprint for this task.

"Chelsea, Toya, and I do not have much work to do in comparison so we can work on UI changes to match the new design." Stephanie offers. Since it is already Tuesday, this story is enough work for A-Team in addition to fixing their gamification bugs for this sprint.

"Great, can Team Sprocket handle a new story and their bugs?"

Paul crosses his arms. "Well, we do have a pretty long list of points to address already but we can try investigating some of those new stories. Usually, we ask for assistance from Team 11 with the backend but if they're busy with deployment, our bugs may take longer."

After reviewing their backlog of bugs with Arash, they agree that there exist a few limitations with viewing lessons which must be addressed since they are fundamental to their platform.

"Would it also make sense to livestream lessons from Jitsi to YouTube so that storage of recorded lessons does not become an issue?" Yazan suggests. Intrigued by the prospect, Arash tells Team 11 and Peter to take on a new spike for recording and replaying lessons in addition to working on deployment, which they may begin implementing in the next sprint in collaboration with Team Sprocket. Meanwhile, Team Rocket will be finalizing their data analytics frontend which they will discuss offline with Arash on how to generate random data to be collected.

Mohammad reminds everyone that the MEng students' final reports for their projects are due on August 24th. The Friday before, the 21st, would ideally be their last meeting, which is in roughly 3 weeks and, ideally, everything should be deployed on the 21st. Thankfully, the undergraduate members are available until the end of August.

Arash confirms with everyone that they are satisfied with meeting their goals one last time and notes that their demo meeting on Friday will only cover their progress.

> **A WORD FROM THE MENTORS**
>
> During this time, deployment of the MVP had higher priority than allocating time towards adopting CICD automation tools. This was a difficult decision for the mentors as it was mainly due to time constraints and lack of experience within the team. As a result, the team did not have an automatic feedback loop from staging to production environments, so there was little room for learning and improving in areas like quality assurance and security. In addition, the team's definition of done was insufficient, as it does not take into account the deployment phase; although the user stories would be considered "done", this wasn't the case. Everything else, however, such as testing, linting, and the GitHub project board, was automated.

Friday, August 7

Since Tuesday, Team Sprocket has worked mainly on view restrictions of certain UI components, with further clarifications from Arash and Mohammad's end regarding existing functionalities as well as "nice-to-have" features.

"I've considered changing some of the hardcoded functionalities for breakout rooms so that a TA may be promoted to instructor for a particular lesson." Paul says.

Mohammad hums. "But the case where an instructor would have an emergency is rare, like if my house catches on fire in the middle of a lesson and a TA would need to continue the lesson for me."

"If it's easy to do, you could look further into this, but the other stories have more priority at the moment." Arash says.

Paul nods in agreement but then stresses. "What about the worst-case scenario where an instructor's internet suddenly dies and no one else has access to the breakout rooms?"

"... I guess we should implement a solution in that case." Arash reluctantly agrees.

Meanwhile, Team Rocket have been receiving help from other teams on fixing some APIs and accessing certain data metrics. Than had expressed her concern with introducing more rework on their end as either having to either create an entirely new backend model or change how their surveys are set up due to the sheer level of difficulty involved with accessing information from the backend. However, with the strategies suggested by other teams, they eventually resolved this issue.

"Will we be incorporating the ignite logo colors?" Stephanie suddenly asks.

"Actually, Mohammad said he was not a fan of orange." Arash shrugs. "In the end, we decided on a shade of blue suggested by our UI designer." He refers the rest of A-Team's more stylistic questions to Mohammad outside of the meeting.

Unfortunately, it seems like something has broken in the backend due to many merge conflicts at the same time. After spending a couple minutes discussing potential sources of this issue, they agree that some members are not able to demo on master today. Instead, they continue planning for the next sprint.

"August 21 is the last demo." Mohammad reminds everyone. "If there's still work left after that deadline, you are responsible for finishing the stories you have picked up."

"We should be able to deploy Jupyterhub by the end of next week." Peter projects optimistically. Nevertheless, Arash urges Team 11 to do as much as they can - even negate working on other stories - to ensure deployment, since issues have continued creeping up on them after starting to work with AWS two weeks ago. With everyone set on their highest priority, the meeting adjourns.

Jerome

The last two days we've primarily worked on setting up the Django backend on AWS and setting up JWT tokens for Jitsi. Both tasks took a lot of trial and error and were primarily accomplished by just trying a bunch of things out and seeing if it would work. We also got a lot of questions from other groups on our live-lesson backend. I think it would have helped if we had made a relationship diagram for our section of the backend that we could have shown people. This would have been very useful if all teams did it from the very start of the project and I believe it would have reduced a lot of confusion. Or if we could have commented our code, we could have explained what each model and viewset does to reduce confusion for teams who have never seen the code before. As our team works full time from 9-5, we have had some issues with communicating or planning zoom calls with the masters students who are asking us questions as they have some classes during the day and often work on the project in the evenings.

Chapter 24: Web Design

Friday, August 14

"The control which designers know in the print medium, and often desire in the web medium, is simply a function of the limitation of the printed page. We should embrace the fact that the web doesn't have the same constraints, and design for this flexibility. But first, we must 'accept the ebb and flow of things."

- John Allsopp

This week, the MEng student teams are primarily focused on finalizing their assigned features as well improving the UI and UX of the Ignite website. The UX features that Team Sprocket is implementing include more quality-of-life features and intuitive steps towards accessing different pages and performing certain actions such as assigning breakout rooms. Team Rocket is now wrapping up on integrating their data analytics frontend to the backend which they have successfully re-implemented, after which they will commence running some tests to generate data.

Having finished nearly all their tasks for the gamification, Team-A+ is now working more on UI restructuring and views of the platform across multiple devices, in close collaboration with Arash and their UI designer through requesting more mockup design versions as well as creating their own along with Masoud's help on certain UI design decisions.

"For making lists responsive on both desktop and mobile, when I test the views for different types of tablets and phones, the behaviors of the styling are different, which makes it more difficult to maintain a uniform view of the site on different devices." Toya explains during one of the standups. Since he won't be able to optimize for all devices, he asks Arash which ones he should prioritize.

After pondering for a couple of seconds, Arash says, "You should just show one column on tables for mobile views and limit the number of characters that you are showing. Depending on the size, then the limit would change, which is usually taken care of by frameworks that expand columns and replace remaining text."

N. Raeesinejad et al., *The Ignite Project*,
https://doi.org/10.1007/978-981-19-4804-6_24

In conclusion, they agree to implement a fixed frame size for mobile view i.e., "small device breakpoint". After discussing and comparing frame sizes for different devices, they also add "medium device breakpoint" and "large device breakpoint". Toya says he'll play around with this some more and consult with Masoud as well.

A WORD FROM THE MENTORS

When aiming for the best user experience, web designers must examine the realities of their users, which is the ultimate factor in determining whether prospects will stick around or click away in frustration! Today, mobile internet users expect web-browsing experience on their phone that is on par with that received on their desktop. This leaves modern web designers with two options: responsive or adaptive design.

A **responsive web website** is responsive to the needs of the users by adjusting its size layout according to the type of device from which it is being browsed. For instance, people visiting the site from a PC or laptop will see the desktop version, whereas those visiting from a mobile device will view a mobile version. A responsive web design is generally initialized to the middle resolution and then uses CSS media queries to change styles based on the target device such as display type, width, height, etc., although only one of such fields is necessary for the site to adapt to different screens. Nonetheless, there will be a few tradeoffs when using media queries since a responsive site may not have a lower loading time than a dedicated mobile site.

Otherwise known as the "progressive enhancement of a website", **adaptive web design** makes use of an established set of multiple static layout sizes to determine the most appropriate size for different types of devices. For instance, when you open a browser on the desktop, the site chooses the best layout for that desktop screen and resizing the browser has no impact on the design. The six common screen widths for an adaptive site are: 320, 480, 760, 960, 1200, and 1600. Adaptive design easily handles correct site rendering; instead of leaving the design to try to rearrange itself for a different screen size, the designers have already picked a different layout. As a result, each of the separate versions of the site may be optimized for a specific device.

There exist several common features between responsive and adaptive web design; Whether the website is accessed through a tiny smartphone or a three-monitor display, all its elements will be properly scaled, well-positioned, and rendered beautifully. Furthermore, both responsive and adaptive web designs are favored by search engines such as Google for improving user experience and promoting easy content consumption.

The boundaries may seem blurred to those without experience in either responsive or adaptive web design, however they differ in certain areas when analyzed closely. The key difference lies in how the design pattern is structured for every device; whereas a responsive design relies on dynamically changing the design pattern to fit the real estate available to it, the content of an adaptive site follows a static layout size. Following proper testing, the latter ensures a fantastic user experience because an adaptive design is made to fit the device. On the other hand, a mobile responsive website is cheaper to maintain; an experienced designer can add a custom responsive code that will give your website the appearance you and your audience will like. Responsive design also takes less work to implement at the cost of opportunity for customization for the design of each screen size. This is most frequently seen with widely available and cheap templates provided by many Content Management Systems (CMS) such as WordPress, Joomla, Wix, etc.

A-Team has almost finished implementing their filesharing service, which encompasses notebook sharing between a student and TA, as well as how the status of notebook questions is shown and recorded. However, they can only test this after deployment.

In the meantime, the undergraduate students are continuing research and work towards connecting and running all services on their servers, such as their "whoami"[36] service on filesharing and oAuth. In addition, they are continuing to figure out how to access stored lesson recordings from their Amazon server.

Today's demo meeting mainly consisted of identifying places of improvement for the UI and UX, such as potential fixes for making certain UI components more dynamic across different devices, adding more "back" and "cancel" buttons to promote user-friendliness, and making a final decision on a uniform color scheme across the site. All the project boards are up to date and clean, with all teams approaching merging everything together.

---

[36] In computing, whoami is a command found on most operating systems and is a concatenation of the words "Who am I?" and prints the effective username of the current user.

"This may look a little silly." Than warns as she prepares to demo Team Rocket's data analytics pages on her screen. "There is not much data as of now, so we created some mock data to show the analytics for."

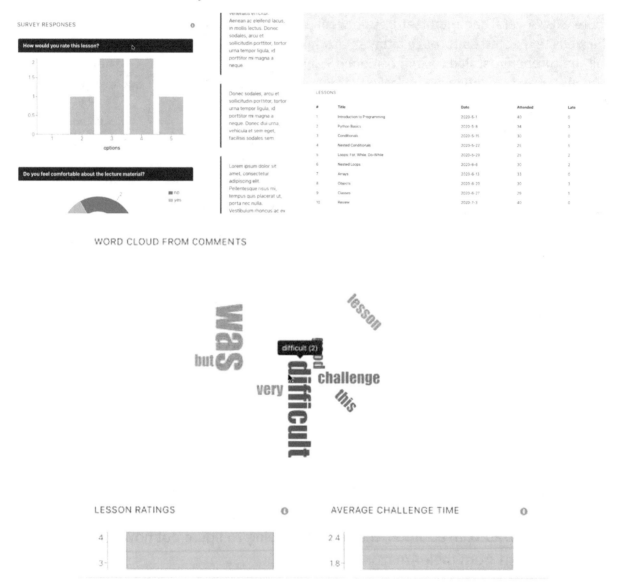

"Please try to filter the words in the word map to only include significant terms." Arash says.

"That won't be a problem to change." Than promises and lastly highlights some additional information that becomes available by hovering over certain data, such as mean and standard deviation of results.

Next, Arash asks them to show how to create a survey and Kevin walks Than through how to demo that on her machine. She first creates an instructor and student user. Then, she starts a challenge by creating and configuring a breakout room. However, after clicking on "Finished Task", something seems to have broken on the website.

"We will investigate this issue later." Kevin assures and proceeds to explain what was supposed to happen instead. The student would be prompted to submit their task, which only records their end-time so they know at what point the student finished the task and each student challenge duration would only count as one entry no matter how many times they submit their challenge. The admin can customize a survey which, for now, is one default template that is offered system wide.

"How big of a change it would be to have the instructor create their own survey?" Arash questions.

"It would take around two or three days to a week."

"Well, everything looks beautiful." Mohammad smiles. "We should also add a disclaimer for students that their data may be used for future research."

Lastly, Team 11 is ready to demo. Yazan shows how a video stream can be recorded and saved as a file in a new folder on an S3 bucket. Their next step is to create a table in the backend to store the S3 bucket so that they can pull up the videos on the frontend side.

Yassin demos their progress towards deploying on Jupyterlab next. After explaining how he got the backend to store data from the frontend, he mentions he needs to keep looking into some inconsistent permission errors that have something to do with the oAuth.

Arash thanks everyone and announces his plans to demo everything in production mode on his own system once they're all done. "Please merge all your changes, fix any migration issues so that starting Tuesday, we can have working development and production environments and so that I can deploy master to production anytime I want from my machine."

"Do we have to test our branches in production mode?" Kevin asks.

Arash shakes his head. "Once Team 11 merges all their changes, everyone will be affected by this. I just want you to fix all your migration issues for master to work properly on Monday."

Monday, August 17

This week, Team 11 will continue working on the frontend i.e., connecting videos and bringing their frontend up on Google Cloud. Yazan will finish the backend today which will be enough for Team Sprocket to go off of for their own tasks. Meanwhile, Yassin will continue working with Peter towards deploying Jupyterhub, which at this point is repeatedly referred to as a slow progress.

"I don't know." Yassin answers sheepishly to Arash's question of when they will be finished with deployment. "Every time we see something in documentation, it seems easy to do, but we keep getting errors whenever we try to implement it."

Arash nods and moves on to list all the items that need to be done before the A-team can test their filesharing feature for any errors. "These items are static IP, filesharing, and the whoami service, which in essence means figuring out how to handle new users. If you guys need more hands on deck, I suggest asking for Bassem and Michael's help as well."

Tuesday, August 18

Team Sprocket made a couple of bugs fixes by adding more buttons. Today they will check in with Team 11 to discuss the look and workflow involved with embedding recorded videos onto the lesson page.

"Production and deployment should be ready today to test everything in those environments." Arash says with a hopeful smile.

"Should we merge before or after production?" Stephanie asks after revealing that Team A+ has submitted a PR for user profile changes.

"Team 11 has submitted a PR for deployment, and we are just waiting for Arash to review it." Yassin notes.

"I may not have time to review it." Arash frowns and allows for Team A+ to merge their changes once their PR is reviewed and accepted.

Wednesday, August 19

"I figured out how to get the video playback to work by embedding the URL in the frontend and I finished the backend for it too." Yazan reports for today's Team 11 standup meeting.

In the meantime, Peter and Yassin have continued working on the Jupyterhub deployment. Although they got past their initial problems with directories not being created,

"Whenever we deploy and try to access service from Jupyterlab, it says service not available which we need to figure out." Yassin says.

"It could be the port or a permission error." Arash suggests.

Peter nods. "Regardless, we'll need your help today to figure this out."

"Also, we cannot access the Ignite website from other machines." Yassin adds with a grimace. "It just refuses to connect."

"Well, it works on my machine." Arash shrugs. "I'll let you know what you may need to change in terms of host settings later."

Thursday, August 20

With the final demo being tomorrow, it seems everybody is nearly done with their tasks. Team Sprocket and Team 11 have successfully gotten recorded lessons to be viewed on the lesson page of the website. With all his tasks done, Yazan will work with Toya on AWS configurations. Peter and Yassin are still working on file permissions and the whoami service, having not resolved all their previous issues yet, and are also looking into implementing Google oAuth.

Team Rocket fixed some bugs and looked into NLP resources to figure out how to filter the word map on their data analytics page.

"We need someone to test our pages but it's very intensive since you would have to generate a lot of data." Than says.

"Even if the others are too busy, we did substantial testing on their own." Kevin adds.

"I will get someone to test them." Arash assures.

A-Team has encountered some difficulties from dealing with several dependencies involved with their CSS modifications for splitting tabs in the frontend, to which they are instructed to remove one by one through trial and error. Meanwhile, Team A+ is working on the course page and user dashboard for both desktop and mobile views.

"I have some ideas of my own as to what to put for text for each tile in the dashboard." Chelsea reveals and with the Mohammad and Arash's blessings, she will implement them in the frontend by tomorrow's demo.

"When the project is done, the only thing left to finish is deployment with Team 11." Mohammad notes.

"At that point, please remember to delete the branches that are merged to master but no longer used." Arash reminds the team.

Mohammad nods and continues. "If there is any glaring bug in the code going forward, we will inform the team responsible for their code. For now, I encourage you all to do as much testing as possible to make sure that your code is perfect."

Toya

The UI design is now complete; Everything is responsive and follows the template from the UI designer. One future task is to add .css files and create script to run upon frontend build. This will override Bulma variables and instead use the Ignite custom colors. Another future task is to figure out how to get authenticated-read to work on S3 buckets. We had made an attempt through AWS Cloudfront to pass signed URLs with expiration dates but this did not work so we will have to keep trying. This project has been very enlightening and provided valuable experience for me. I feel that I become very strong in frontend development, but I also feel like I am also fairly proficient in backend development and cloud service deployment now. I still don't believe I am anywhere close to being skilled in full stack development, but this project has shown me many areas in software that I could pursue. Now that I'm done, I will continue to job hunt, but also take online courses in cloud services, 3D game development (Unity), and continue to work on my frontend / backend skills.

Friday, August 21

Today marks the last official demo meeting with all the teams present.

Team 11 has finished the backend for re-watching lesson recordings and helped Team Sprocket with connecting it to the frontend. They have also fixed some bugs involved with emails and unauthorized blacklists. Peter and Yassin have moved on to implementing the backend for their whoami service, however they still need to figure out if authorization is working I.e., if cookies are being sent properly.

"Will we do any integration testing for the whole system once everything is deployed?" Mihai asks.

Arash nods in agreement. "Hopefully we'll get to do that next week once deployment is done and we can log in to the system."

According to Yassin, the deployment is still not demo-able due to authentication errors and the fact that their working branch is a couple days behind master. As a result, Arash asks everyone to demo from their own machines. After a couple of minutes used to set up their systems, Team Sprocket kicks of the demo.

"Going live and splitting tabs during live lessons has been optimized for smaller aspect ratios in mobile views." Paul explains as he navigates through different pages on his shared screen. "The functionalities are mainly the same as before; only the UX is different. Navigation is easier and more intuitive. We also added some backend changes for embedding lesson recordings."

Toya clears his throat "Everything is now viewable on mobile and desktop. Bulma CSS does not work entirely correctly in Jupyterlab, so we had a chat with Masoud wo work around this issue."

"When building an app with React, it does not assume that you have tabs within tabs which creates some problems." Paul adds. "For example, Team Sprocket experienced some issues with the size of modals. I think the best way to address this is to not gave modals where you need to do scrolling, we should be considered for the layout in the future."

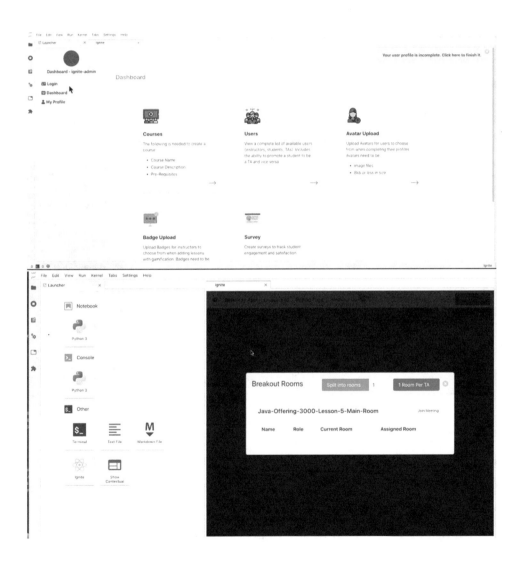

"The lesson page looks different from what I expected." Arash admits. "The components seem to be all over the place."

"I tried to fix the columns for different views through different methods which didn't work." Stephanie says sheepishly.

"It might be better to have everything centered in the desktop view similar to a mobile view."

"That would look nice." Mohammad agrees.

Stephanie nods and then shrugs. "I was just trying to replicate the UI designs as much as I could."

"How hard would it be to change the design to what we want?"

"It may not be too bad." Toya answers. "but it would require a couple hundred lines of code to be changed." In conclusion, they agree to change the content of the page to be centered and surrounded with more whitespace.

Lastly, Team A+ quickly show off their finished upload page for avatars and badges which have been implemented to look identical to their provided UI designs.

"Awesome job everyone!" Arash congratulates the team. "I hope you have learned a thing or two. Hopefully it was a fruitful experience and I wish you the best of luck in your future endeavors."

"The Ignite Club will want a demo of the website soon." Mohammad adds with a bright grin. "Truly, a job well done to all of you!"

Chelsea

Today was the last demo...It was bittersweet because there really wasn't much left to demo but we got to unveil our styling changes to everyone. There were some things that we were unable to make look exactly like the UI design. That may be because we chose to stick to Bulma, but overall, I think we matched it really well and got the mobile responsiveness to be as seamless as possible. There are still some things that could maybe look better, but I honestly cannot see students trying to manage multiple tabs on something like a smart phone and that is where the responsiveness has some weaknesses. It's going to be strange to not be working on Ignite anymore, though I am really excited to see and experience what happens next!

Chapter 25: The Stakeholders

Friday, August 28

"If there is a worse time for something to go wrong, it will happen then."

- *Murphy's general laws*

Mihai

This week was the final week of capstone and working on the Ignite project. Thomas, Paul, and I managed to finish embedding recorded videos into the Lesson page with help from Yazan and Jerome of Team 11. We mopped up the last of the bugs and finished up the last of our tasks in time for the demo.

Looking back on it all, I think this capstone project made my experience in this program much richer. Besides learning React and a few other things on the technical side, I learned what it was like to work on a REAL project in a BIG team - that kind of experience is much harder to come by. There were definitely challenges along the way but it's amazing to see what teamwork can accomplish. Sometimes just having another person stare at the code with you leads to some breakthrough. Arash, Masoud, Yousef, and everyone else that supported us were very helpful and I thank them for their help.

Looking forward, I definitely want to learn a bit more about backend work and devops as these things were the biggest technical challenges. I will also play around with docker a bit more and see if I can find some good resources on authentication. It'll be interesting to hear what the first wave of users think of Ignite and how it could be improved!

N. Raeesinejad et al., *The Ignite Project*,
https://doi.org/10.1007/978-981-19-4804-6_25

Last week, I had the opportunity to listen to the MEng students' final presentations for their capstone course. In addition to their work scope and methodologies, each team discussed individual challenges that they faced, some of which were common between all teams, such as adapting to new tools and environments, which came to no surprise to me. I was more intrigued, however, by the lessons they learned and recommendations for future work on this project. It was evident that everyone was truly invested in successfully deploying the Ignite platform, as they continued to improve their features as well as deal with error-handling well after the official deployment deadline.

Team-A, for instance, reported that their biggest remaining tasks involve modifying connections between their backend and individual services once they stop using localhost and assisting the deployment team. In contrast, Team Sprocket will mainly focus on refactoring their existing codebase, implementing UX improvements based on feedback, and making the website more mobile-friendly.

Another hot topic discussed during presentations was reliability; Mohammad asked every team how much testing they have conducted so far. While most teams have tested their features on different machines, they did not manage to test certain services for different users; these services would require substantial testing during deployment and on a matter of scale.

Today marks the last meeting with all the teams. While feeling bittersweet about the end of this project, I am nevertheless very eager to see the deployed MVP. Arash has even encouraged Mohammad to invite the Schulich Ignite mentors to attend as well, since they are also stakeholders who will be using the platform to hold interactive coding workshops with high school students in the coming Fall semester.

A WORD FROM THE MENTORS

A stakeholder may refer to a person, group, or organization that affects or can be affected by an organization's actions. External stakeholders are typically external to the organization that is developing a product for customers, partners, and regulators. Alternatively, internal stakeholders are internal to the organization that is developing the product for senior executives, managers, and internal users.

One of the most important feedback loops in agile is sprint review meetings. As such, stakeholders not participating in sprint reviews often leads to misunderstandings being caught later in the development process, sometimes so late that deadlines shift. Surprising stakeholders generally leads to disappointment and frustration, which can be avoided by soliciting early and frequent feedback as well as sharing information transparently.

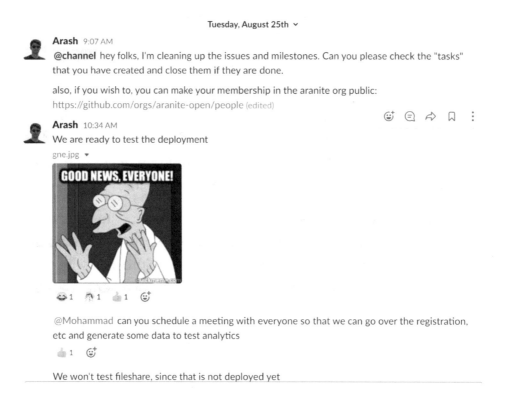

"The goal today is to figure out any problems we may see." Arash starts. "There may be some bugs and issues. If you see something that is blocking you and you can't proceed, please let me know and take note of anything you find that may need improvement. Expect some problems because I don't expect everything to go smoothly."

"The executives of Ignite are our stakeholders and will make precise notes of bugs and suggestions for the system," Mohammad announces. "They will have a key role going forward by discussing the features and bugs in focus groups as well as training other mentors on how to use the system."

The current Schulich Ignite president smiles and waves. "We're looking forward to it."

For the initial setup, Arash assumes the role of system admin and Mohammad acts as the instructor on the website. Meanwhile, the rest of the team and Ignite mentors will login to the system as students.

Right off the bat, the first problem that a couple participants encounter is a "server not found" error upon entering the website URL and Arash tells them to not use the URL they were using for deployment. After Peter also notes that his chrome browser ran into a 503 error, they spend the next few minutes changing their hosts to be directed to the website.

After finally accessing the homepage, some Ignite mentors speak up about having trouble navigating to the profile page, noting that student users who would be beginners to programming will most likely make little to no sense of the meaning of the tiles in the launcher page.

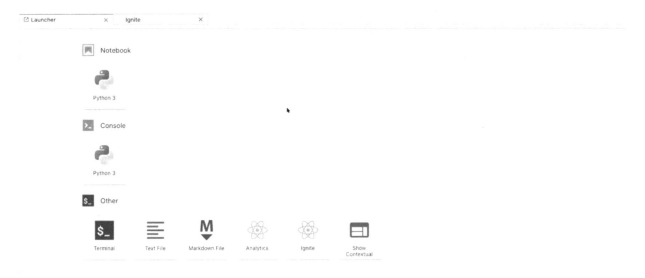

Arash tells them to click on the "Ignite" page for the time being. "This launcher is a placeholder for now and will be removed later so that the Ignite page will be the homepage of the site."

With further suggestions from the focus group, they ultimately decide for students to only be able to view the Python notebook, console, and Ignite icons on the launch page, along with additional help text to navigate between pages.

Immediately afterwards, a couple more people report receiving a "400: bad request error".

"Looks like our setup is not good enough for this number of people on the site." Arash frowns.

"How many nodes are we running on?" Peter asks. Navigating to their Google Cloud Platform instance named "ignite-aranite-pool", Arash reports that they are currently running their service on three nodes.

"The maximum amount you can have is eight."

"In that case, let's change it from a three to an eight to try to support the number of users currently on the site." Arash concludes and edits their node pool size accordingly. "With three nodes, we were able to support twelve users."

Mohammad nods. "This was all part of our load testing. Proportionally, we will need to scale this up for one to two hundred students."

A couple of minutes pass in silence as the cluster starts to rebuild.

"I am encountering a timeout error." Than suddenly remarks and Arash tells her to hold shift and refresh her browser to clear the cache. Once this error is resolved, she then reports another error stating that their server is not running, which other Ignite mentors echo in the chat.

"Please wait until the cluster finishes rebuilding with the updated number of nodes." Arash tells everyone. "It will take around fifteen minutes to rebuild." In the meantime, he requests those who have already logged in to set up their profiles.

"Once we get the system up and running and we can actually operate, the Ignite executives and focus groups will build a list of requirements for us to add to user profiles." Mohammad notes.

As the admin, Arash proceeds to create a new survey with a variety of question types such as scale, binary, and open field to generate some data for data analytics, informing the Ignite mentors that they can customize their own brief surveys after their lessons.

Next, he creates a new course with the title "Ignite 101" and demonstrates how a faculty user may add new course offerings and lessons. After creating two offerings – one draft and another published – he enrolls a list of students by uploading a CSV file containing the emails of the meeting participants. The stakeholders and the team note some more non-intuitive UI and UX features for improvement along the way, such as size scaling for different devices, compatibility with different email formats, and accessibility criteria for different types of users.

They spend the next hour simulating a live lesson, continuing their conversation over a Jitsi meeting on the system. With the end of the simulation, the team conclude their demo by sharing some final comments.

"Everything involved with creating lessons is pretty manageable with minor changes." Mohammad starts. "A couple of the services that are supposed to communicate with each other will need some ironing out and there were some bugs like students being able to create breakout rooms during a live lesson. After fixing these issues, we can focus more on the user experience. Otherwise, if the system's connection does not break, it should run relatively smoothly."

"We should also juice up the Jitsi video servers." Yazan adds. "This shouldn't be too difficult; we'll just need to stop all the servers, scale them up, and then start and test them all over again. We'll also need to implement more video bridges to manage the load, because people kept getting kicked out and sent back to the lesson."

Mohammad nods. "Exactly. The framework is there, we just need to make sure that it is scalable and can handle the load prior to deploying. This should be our main priority since it was the root of most of our issues along with some minor glitches."

"Will we be receiving documentation for how to use this system?" The Ignite Club President asks.

"Well, the system is a moving target as we continue to rapidly change things per the suggestions of the stakeholders." Mohammad explains. "It will be fairly costly to constantly change our documentation accordingly. Perhaps once the system becomes more stable, we can create some documentation. Another thing we can do is make the system more self-explanatory through help buttons and text all over the website. At the end of the day, we want the system to be intuitive enough that you wouldn't really need a manual to work with it."

Arash hums in agreement. "And mind you, an instructor such as yourselves could easily work around some clunky features. Our main concern is making the system easy-to-use for students, since there will be around two hundred of them enrolled."

Concluding the end of their deployment testing and demo, they spend the remainder of the meeting discussing how to better scale the Ignite System with additional input from the Ignite club executives. Finally, the stakeholders thank the team for their hard work in developing this platform and Arash ends the meeting with a tired but pleased expression.

Epilogue

The President of Schulich Ignite

The Ignite Project had a significant influence on how the Schulich Ignite workshops were delivered to high school students and positively enhanced the user experience for both mentors and mentees in the Ignite program. The Aranite platform was utilized to host coding workshops for high school students as part of the Schulich Ignite initiative, which aims to make coding and programming principles more accessible to high school students. The Aranite platform allowed mentees to code with confidence alongside allowing mentors to teach with ease as a result of the simplicity, convenience, and clarity the platform provided. The mentees in the program had a favorable user experience since there were no download or installation processes for students to undertake; instead, it was a simple online link that students could follow and further access the platform by logging in with their Gmail accounts. This flexibility allowed mentors to concentrate on teaching students the learning materials rather than troubleshooting complicated installation processes, which usually occupied valuable time during Ignite sessions. The utilization of this platform benefitted numerous students as accessibility to the platform did not necessitate the use of highly advanced computers to achieve a successful setup of the coding environments. Furthermore, the platform enabled students to code directly on the browser and save their work locally or on the platform, which was extremely convenient for all Ignite participants for project development.

Many students were able to use the Aranite platform to create complex projects as it provided great functionality and compatibility for software development. This had a significant beneficial influence on students' learning as they were able to leverage their fundamental coding knowledge to create innovative and challenging games using the platform. As a result, the platform eliminated the complex hurdles that beginners and/or young students experience while learning to code by giving a single access IDE right on the browser with additional valuable user functions that further enhanced the user experience. The documentation offered by Aranite was an excellent resource for new users to become comfortable with the platform and feel confident in its operation, which further assisted students in their learning.

As per future work for this project, the platform can integrate potential enhancements to support more users concurrently while preserving proper and thorough access for all users. This can make it possible for many mentors and mentees to use the platform simultaneously without encountering any delays. Additionally, the platform's incorporation of more robust and compatible video conferencing technology can further streamline the online learning process by allowing mentors and mentees to connect and code on the same platform rather than requiring them to utilize separate video conferencing software. Overall, the platform was incredibly great for Schulich Ignite since it made learning to code accessible and simpler for students with little to no coding experience. As a result of this innovative initiative, the Schulich Ignite program greatly benefited as sessions were more successful and geared toward teaching

N. Raeesinejad et al., *The Ignite Project*,
https://doi.org/10.1007/978-981-19-4804-6

coding principles rather than dealing with complicated software, allowing mentees more time and resources towards leveraging their core coding knowledge and skills. This project truly serves great potential and is one that is greatly beneficial for beginners or anyone looking to get started with coding as it easily assists students in leveraging their coding knowledge.

Niyousha

It has been almost 2 years since I attended the final demo of the Ignite system. Although I was concerned with not being able to get hands-on experience on the project itself in the beginning, I had the opportunity to gain valuable insights into software engineering industry best practices as a close observer of the process followed by the Ignite team as they were building the MVP over the course of roughly 5 months. Prior to the project, as a software engineering student who had just finished her second year, I had no notion of Scrum; the concept was entirely new to me. However, I quickly noticed the benefit of integrating the agile framework with the MEng and PURE student teams. It was evident that it not only encouraged the students to self-organize when working on issues, but to also reflect on their successes and challenges, and prioritize learning throughout the entire experience.

Before long, I began to apply the principles of agile myself, particularly in other areas of my academic and extracurricular activities. For example, a couple months after the final demo of the Ignite system MVP, I started my 3rd year in software engineering and decided to join Tech Start UCalgary, a student driven club which focuses on software development and entrepreneurship at the University of Calgary. As a developer on the club's website team, it was important for me to help establish a dynamic and collaborative workflow for my team so that we could be effective with our learning and develop a top-quality website by the end of the school semester. So, I created my own guide and delivered a talk on Agile web development to my team, based on all the concepts I had learned during the Ignite team's Scrum meetings. The concepts I introduced were relatively abstract to everyone on my team, which was expected since none of our members had any practical software development experience as 1st to 3rd year students. Also, as someone who only observed and had theoretical knowledge of the agile framework, there was little I could do to guide my team like a Scrum master. Although we ended up building an exceptional website for our club, there were many cases of miscommunication and code smells as we rushed to develop a working website before the end of the Winter semester.[37]

The following Spring, I finally participated in my first standup as I started my 16-month software engineering internship at Pason Systems, an energy services and technology company. I discovered that agile teams are largely self-governing, and each team may adopt the agile methodology differently to suit their business and development needs, such as having different time durations for sprints, or a different sequence of effort estimation values for planning poker. It was a new experience for me to participate in only my own team's standups, in contrast to the one collective standup that the three Ignite teams attended. Furthermore, due to all the industry insights that I had gained from the Ignite project, I found that my onboarding process was very quick, and smoother compared to the other new hires at my company, and I was able to fully participate in our standup, sprint planning, and retrospective meetings from the beginning. I also found that my training was significantly less time-constrained compared to that of the Ignite team, because the duration of my internship allowed for it as opposed to the 4-month deadline for the Ignite system MVP as part of the MEng course for Team Design Project in Software Engineering.

When I returned the following Fall semester as a Project Manager of the website enhancement and maintenance team for Tech Start UCalgary, I took on a more practical approach when getting my team adopt the Agile development workflow for the first time. My main goals with this approach were to better streamline our development efforts, promote more communication within the team, and encourage everyone to review and provide feedback for each other's work. The biggest lesson I have learned from this project management experience is to invest more time and provide more just-in-time and one-on-one support during the early project stages so that the members become more knowledgeable and confident in their technical skills before being assigned to more involved tasks such as spikes and design discussions, like the support from the mentors for the Ignite team, in form of just in time coding workshops related to their issues and pair programming. What I discovered was that the students on my team – namely those who had not yet started their internships – were appreciative of gaining practical experience following a development process that is reflective of industry. As a result, I was able to be a successful project manager in ensuring effective communication within my team and help them stay on top of their issues despite being full time students, which is largely attributed to the agile methodology and software engineering best practices that I was exposed to throughout my experience with the Ignite project.

---

[37] https://techstartucalgary.com/

Mohammad

The idea for the Ignite Project was born out of a simple but fundamental teaching philosophy:

Software Engineering is a practical profession, and software engineering education must follow suit!

Software engineers are practitioners who are often tasked with the design, development, and maintenance of complex systems. An effective software engineering career that prepares students with a career in tech can greatly increase their productivity in their starting years and improve their job satisfaction and the longevity of their careers.

The fast-paced and ever-evolving nature of technology demands software engineers to become accustomed to change, self-learning, and simply keeping up with an influx of tools, frameworks, methodologies, etc.

The skills of software engineers are trusted upon a strong basis of fundamental technical knowledge such as data structures and algorithms, various programming paradigms, distributed systems, computer architecture, etc. This fundamental technical knowledge makes up the vast majority of what is traditionally taught by universities and educational institutions.

In addition to fundamental and advanced technical knowledge, software engineers must possess soft skills including communication, teamwork, and presentation skills. Software projects are almost always done in teams, often dealing with multiple stakeholders with different backgrounds. In addition to teamwork and communication skills, software engineers must have a practical understanding of software development process methodologies such as scrum.

Our teaching philosophy led us to an important question: How can we effectively train software engineers for the dynamic challenges of the industry?

The direct stakeholders impacted by this question are:

- Software engineering students, interns, and new graduates
- Academics, educators, and educational institutions
- Experienced industry engineers, team leads, and engineering managers, but also experienced engineers who will work with interns and new graduates and will inevitably contribute to their training.

The indirect stakeholders of this question are anyone who uses software in the world!

Consequently, the three groups identified above are the target audience for this book.

As an educator, throughout the ignite project, I experienced being the invisible man in the room, or the fly on the wall, and observing my students' first day at work. It was exciting, nerve-wracking, and at the same time, illuminating. I knew my students are good programmers and already have a practical understanding of software development best practices. I also knew they were all relatively new to web development and had limited exposure to cloud computing and deployment, as well as a limited understanding of agile methodologies. Therefore, it was not a surprise to see some additional on-the-job training is required. As we observed the students work through their first couple of weeks to a month of their internship, it became apparent that many of the challenges involved applied and soft skills. While the students were fairly proficient in self-learning, it was apparent that they are not accustomed to the industry environment and project setups. And although the students were able to make the required adjustments in the weeks to come, the question remains: what can we do as educators to minimize the adjustments needed?

We suspect the answer resides in a healthy balance of the three above-mentioned categories of fundamental knowledge, applied skills, and soft skills, combined with a student-centric pedagogy that promotes active self-learning and creative problem-solving.

As an educator, some of my key practical takeaways include:

- ✓ Coupling of strong fundamental knowledge and self-learning ability – While technology consistently changes, the fundamental concepts of computer science and software engineering experience little alterations. Students greatly benefit from detailed lectures and instructions on key topics such as algorithms, memory management, etc. However, student-centric pedagogical approaches that promote self-learning are key for software engineers as it prepares them for a career that demands consistent learning.

✓ The importance of mentorship and just-in-time support – Adaptive learning and modeling are of paramount importance they help students find their path efficiently while powering them to be self-reliant in their training.

✓ Early Exposure to Industry Practices and Experience Working with Industry – Project-based learning is a great start as a student-centric learning activity. However, to maximize learning for software engineers it must be coupled with agile development, software engineering tools, and if possible, industry-based projects.

For students, the Ignite Project will serve as a roadmap where they can read the story of this internship, to grasp an understanding of their start in the industry. Students can take inventory of their current skill level and develop an understanding of the path ahead.

By the end of the summer, we had created real value in our software engineering team! Our group of interns was now well-trained and actively contributing to the project at a fairly high rate. We suspect this type of effective value generation will be of great interest to the industry. The pedagogical approaches used to support the learning of software engineering interns in the ignite project and the mentorship structure can greatly enhance the work experience of software engineering interns and new graduates.

The training of the software engineering interns in the Ignite project was a collaborative effort between academics, the industry, and the students. The academic and industry advisors discussed fundamental and applied approaches to student training and worked to find agile solutions to maximize impact while learning from the feedback and progress of the students.

And thus, we present the Ignite Project to you, with another simple but fundamental idea:

Software Engineering is a collaborative profession, and software engineering education must follow suit!

The End

Printed in the United States
by Baker & Taylor Publisher Services